Cooperation Management for Practitioners

GIZ GmbH

Cooperation Management for Practitioners

Managing Social Change
with Capacity WORKS

 Springer Gabler

GIZ GmbH
Deutsche Gesellschaft für internationale Zusammenarbeit
Eschborn, Germany

ISBN 978-3-658-07904-0
ISBN 978-3-658-07905-5 (eBook)

The Deutsche Nationalbibliothek lists this publication in der Deutsche Nationalbibliografie; detailed bibliographic data are available in the internet at http://dnb.d-nb.de.

Springer Gabler
© Springer Fachmedien Wiesbaden 2015

Springer-Gabler is a trademark of Springer DE. Springer DE is part of Springer Science+Business Media.
www.springer-gabler.de

Foreword

Successful cooperation is a key challenge of the 21st century. Be it health reform or airport expansion, education system reform or public debt reduction, neighbourhood development or the integration of migrants: in Europe and worldwide, these and similar change processes are now on the agenda more than ever before. Crucial to their success is not just selecting the right political and technical solutions, but also the issue of how to transform them step by step into a new social reality. This means finding the right methodology.

Through this book we are now making available the quintessence of our experience in this field, gained in over 30 years of international cooperation by GIZ and its predecessor organisations GTZ, the German Development Service and Capacity Development International, Germany. Through our management model Capacity WORKS we are showing what we believe makes cooperation succeed, so that others can benefit from this expertise.

Organisations rarely know just how much they actually know. What is especially valuable, though, is their wealth of experience, the methods of the seasoned practitioner, and an understanding of the principles underlying this know-how. This knowledge is reflected in the practices of the knowledge bearers, and is transmitted orally. Yet it often remains implicit. So to make our expertise available for wider use – both within GIZ and by others – we needed to decode, condense and express it in a way that is easy to understand. This is why we embarked on a shared voyage of discovery – a journey to the factors for successfully managing social change. We wanted to know what it actually is that makes those projects which generate results particularly effectively and sustainably, better than others. To answer this question we collated the lessons our practitioners had learned, and analysed them in light of recent systems theory.

The product of this analysis and reflection work is the management model Capacity WORKS. Capacity WORKS is a key tool that supports us and our partners around the world every day in our work for social change processes. In diffuse and complex constellations, Capacity WORKS gives the actors involved guidance and structure, without constraining them. And using simple methods it facilitates a joint understanding of the key issues in the joint project and how to approach them. But that is not all. At the same time Capacity WORKS also reflects an attitude, and articulates a standard of quality for cooperation projects. The distinguishing feature of these projects is that all the actors involved participate actively, listen and look carefully, pool and negotiate their interests and strengths, and continuously reflect on their joint undertaking.

'Cooperation management for practitioners – Managing social change with Capacity WORKS' is a manual for GIZ staff and partners worldwide. But it is also designed for everyone involved in cooperation systems on any level whatsoever – whether as managers, executives, consultants or advisors in business, governance, public administration or the nonprofit sector. We hope that you will find it helpful in driving key reforms and change processes. Good luck!

Dr. Christoph Beier
Vice-Chair of the Management Board

Cornelia Richter
Managing Director

Deutsche Gesellschaft für Internationale Zusammenarbeit (GIZ) GmbH

Contents

Toolbox

Toolbox | Success factor Steering structure 182

Toolbox | Success factor Processes 205

Toolbox | Success factor Learning & innovation 225

Explanation of symbols

 GIZ-specific    Practical example

Introduction

Cooperation is the cornerstone of social development, no matter where in the world. People create societies through cooperation, and no actor can manage this process on their own. They can only achieve this through good cooperation relationships at the local level, within entire societies and increasingly across national borders. The days when only states and governments cooperated with each other are long gone. Civil society and private sector actors are now also joining cooperation systems, to help develop joint responses to urgent issues such as sustainable energy supply and climate change.

As familiar as the phenomenon of cooperation may be, the actual task of managing it proves no less complex. Long ago, Goethe wrote in *Faust*: 'Think about the What and, even more, the How!' Each actor involved brings with them their very own understanding of this 'what'. If cooperation is to succeed, the very first thing we need to do is find a way of turning these many different ideas into a shared understanding of the 'what'. But then it is not yet clear how we should go about reaching this 'what'. In other words, we have to define the 'how', i.e. the specific steps and interventions that will take us to our joint objective.

In 2006, these insights led us to begin developing Capacity WORKS. We answered the question of how to make cooperation a success not from a theoretical perspective, but on the basis of practical experience with international cooperation. We asked experienced consultants what their 'formulas' for successful cooperation would be. Their answers varied as widely as the cooperation systems from which we were seeking to draw lessons. One thing did become clear, though: organisations will only participate in a cooperation system, and allow themselves to become dependent on others, if they find the objectives attractive and cannot achieve them on their own. We asked how this works. In the answers we received, certain patterns kept repeating themselves. We identified these as 'success factors' for professional cooperation management:

- Strategy: The cooperation system will succeed if and when the cooperation partners agree on a joint strategy to achieve the negotiated objectives.

- Cooperation: Trust, the negotiation of appropriate forms of cooperation, and clearly defined roles form the basis for good cooperation.

- Steering structure: The cooperation system is guided by agreements on how the actors involved will go about jointly preparing and taking the decisions that affect them.

- Processes: Successful cooperation systems include a clear understanding of effective ways of delivering outputs, for which new processes are established or existing processes modified.

- Learning & innovation: The cooperation partners create an enabling environment for innovation by boosting the learning capacities of the actors involved.

As they manage this process, the cooperation partners switch back and forth between observing and analysing their setting, and implementing concrete actions designed to facilitate change.

Capacity WORKS describes all of this, and provides a mature methodology for the art of successful cooperation management using relevant key questions, the structure of the five success factors, and a toolbox for addressing specific questions.

These elements combine to form a management model for analysing and understanding complex cooperation systems in a structured way, without offering blueprints. The model supports cooperation partners in developing a common language to articulate 'what' they wish to achieve, and 'how' they intend to achieve it by cooperating. It helps orient cooperation management systematically toward the objectives and results that they wish to achieve. The partners involved thus attempt to jointly interpret their reality and map out a desirable future. This enables them to take joint decisions, and implement these as effectively and efficiently as possible. Capacity WORKS does not provide any ready-made answers to the particular challenges of each and every cooperation system – just as 'painting by numbers' will not turn any of us into true artists.

Capacity WORKS is designed for anyone working in or advising complex cooperation systems. It builds on the current professional discourse on cooperation and change management, and at the same time reflects the long-standing experience of the Deutsche Gesellschaft für Internationale Zusammenarbeit (GIZ) GmbH.

Capacity WORKS – a management model in international cooperation

Capacity WORKS has proved a great success in German international cooperation, where it has achieved an excellent track record as a model for cooperation management. This has been the case regardless of the considerable differences between the various countries, cultures and sectors in which we work. Capacity WORKS was first developed in 2006 within the former GTZ, one of the predecessor organisations of today's GIZ. Following a two-year pilot phase in 2009 and 2010 we then introduced it as our management model for sustainable development. Today, Capacity WORKS is an integral part of all GIZ's key procedures from programme design, to implementation, to internal evaluation and reporting.

GIZ – a federal enterprise

GIZ is a German federal enterprise. Most of its work is commissioned by the German Federal Ministry for Economic Cooperation and Development (BMZ). The company also operates on behalf of other German ministries, as well as public and private sector clients in Germany and abroad. These include the European Commission, the United Nations, the World Bank, governments of other countries, the Bill & Melinda Gates Foundation, private businesses and German states. GIZ maintains a presence in more than 130 countries worldwide. In Germany, we maintain a presence in nearly all the federal states. When GIZ emerged from the fusion of the German Development Service (DED), the Deutsche Gesellschaft für Technische Zusammenarbeit (GTZ) GmbH and InWEnt (Capacity Building International, Germany) in 2011, the German Government defined an expanded corporate purpose for the new company. This refers to 'promoting international cooperation for sustainable development'. The common element in all GIZ's services remains the promotion of global sustainability to protect and preserve the future of the South and the North. The expansion of GIZ's corporate purpose enables the company to broaden its work both in Germany and for EU member states and emerging economies.

Projects

To understand the examples of GIZ projects described in the present manual it is important to be familiar with a few terms and to know how German international cooperation works. The term 'project' refers to projects (or programmes) that aim to support social, political and/or economic change within a defined time frame. Most of GIZ's business involves projects commissioned by the German Federal Ministry for Economic Cooperation and Development (BMZ). Responsibility for implementing these is shared between the partner country in question and GIZ.

In these cases the decision to engage in a partner country rests with BMZ, and is agreed with that country in the policy dialogue. GIZ supports the preparation of projects, but is not commissioned to implement them until a concrete implementation proposal has been put forward. So regarding the 'what' of cooperation there are always certain directives – usually BMZ (or other federal ministries) support change processes or reforms which the partner countries are already working to achieve. The 'how' of cooperation is then negotiated between the government actors and GIZ project managers in the partner countries concerned. GIZ is accountable to BMZ (or other commissioning party) for ensuring that the objectives and results agreed in the commission are achieved. At the same time, GIZ works jointly with the actors involved in the project to achieve the agreed objectives and results. These are integrated into the strategies, policies and programmes of the partner countries, and harmonised with the contributions provided by other international actors.

Areas of social concern

In most cases the present manual uses the following terms synonymously: 'sector', 'policy field', 'social sub-system', 'area of social concern', 'social context', 'permanent cooperation system'. These terms always denote the area or sphere to which the objectives of a project relate and in which change is to be brought about. In many cases the envisaged changes to be facilitated through projects relate to specific sectors that in most governments are represented by line ministries. Health, education, public administration, agriculture and water supply are instances of such clearly defined sub-systems. In other cases, the changes affect a number of sectors. Adaptation to climate change, youth protection or rural development, for instance, often require changes in many social sub-systems. If we define the area of social concern as 'tackling the impacts of climate change', for example, then water, agriculture and business are all important sectors in which a change process might be facilitated. The term 'areas of social concern' enables us to refer to several sectors at the same time, thus avoiding a too-narrow sectoral perspective on change processes.

Capacity WORKS was originally developed for projects focusing on the capacity development of partner countries as understood in the context of technical cooperation. In this context, 'capacity' means the ability of people, organisations and societies to manage their own sustainable development processes and adapt to changing circumstances. This includes recognising obstacles to development, designing strategies to tackle them, and then successfully implementing these.

At GIZ this ability is often described as 'proactive management capacity'. This ability encompasses the political will, interests, knowledge, values and financial resources that the agents concerned need in order to achieve their own development goals. In other words, capacity development is about developing capacities of individuals, organisations and societies so that partner countries can articulate, negotiate and realise the processes of reform and development that they themselves envisage.

GIZ (and its predecessor organisations) possess more than 30 years of experience with this core competence. Today, alongside 'traditional' capacity development services more and more projects are emerging that reflect the needs and goals of other clients, and require new ways of approaching them. Examples of this include management and logistical services, advisory services for global partnerships and fund management. Regardless of the client or clients involved, all these services are subsumed under the term 'international cooperation'. The Capacity WORKS management model has always proved expedient in situations where cooperation systems arise. It supports users in managing the processes of negotiation and decision-making involving various actors, so that the desired objectives and results are achieved sustainably.

Sustainable development – the guiding principle

GIZ is guided by the principle of sustainability. We see sustainable development as the interplay of social responsibility, ecological balance, political participation and economic capability. Only this will enable present and future generations to secure a life of dignity. The way we work reflects this principle in all areas. The various dimensions of this understanding of sustainability are inherent in the management model Capacity WORKS.

The goals of economic development, social justice, environmental integrity and political participation often conflict. Viable solutions to these conflicts need to be found in social, cultural and political contexts that are constantly shifting. Issues of power and interests play a major role here. Sustainability therefore requires a permanent, ongoing process of negotiation so that workable compromises can be reached. The state, the private sector and civil society are all affected, and must all be involved in striking these compromises at all levels – locally, nationally, regionally and internationally.

By involving the relevant actors the management model Capacity WORKS enables users to develop sustainable solutions that are tailor-made to suit the given context. This is achieved most effectively by making transparent and negotiating the different perspectives on any given issue.

Managing cooperation for global solutions

Integrating different perspectives is especially important when solving transboundary or global problems. Climate change, fair trade regimes and financial market stability are examples of global challenges that require a concerted approach.

International cooperation can be made significantly more effective when the relevant actors systematically share knowledge and experience regarding their competencies, perspectives and lessons learned. Cooperation systems are formed around specific issues, and develop joint problem-solving approaches. Approaches that have enabled a society to successfully tackle a specific issue can also be made accessible for other countries. Sharing success stories creates access to relevant implicit knowledge, and can be used to jointly develop innovations ('co-creation').

This kind of knowledge sharing is becoming more and more important as we move toward a global knowledge, information and media society. Here too the guiding principle of sustainable development should play a central role, in order to facilitate generational justice and political participation. GIZ already uses these learning processes when supporting cooperation between developing countries, as well as between emerging economies, developing countries and industrialised countries. These international processes are particularly challenging for cooperation management, however. Here too the logic, structure and tools of Capacity WORKS can help support the sustainability of the results of these cooperation arrangements.

What's new in Capacity WORKS?

As we mentioned at the beginning, using Capacity WORKS means looking explicitly at the 'what' and the 'how' of managing cooperation systems. GIZ has benefited from describing how it works and being able to refer to a conceptually and theoretically sound model.

Capacity WORKS is a storehouse of the company's management experience gained in 30 years of international cooperation. Just how important this step was, is underlined by the wealth of positive user feedback at GIZ. The ultimate yardstick of the model's success is its usefulness to those involved in continuously seeking solutions for cooperation systems.

The merger of GIZ's predecessor organisations DED, GTZ and Capacity Building International, Germany provided an opportunity to also make this cooperation expertise available to those parts of the company that had not been involved in developing it. At the same time, new experiences invariably teach us new lessons. As Capacity WORKS was gradually rolled out at GIZ, the ideas contained in the success factors were fine-tuned, and the contexts in which they are applied were broadened. These developments made it necessary to update the management model. This new version makes available the state of the art in cooperation management at GIZ.

Existing users will notice that this version of Capacity WORKS looks much more closely at the topic of objectives and results. The distinction drawn between cooperation systems and networks also adds an important and enriching dimension to the model. The new version describes with even greater clarity the management of cross-organisational strategy development. Learning and innovation processes are examined with a stronger focus on knowledge sharing. All these new developments are also reflected in the modified toolbox, which supports concrete managerial decision-making in cooperation systems. The toolbox contains both tried and tested tools, and new ones. It contains methodologies that are particularly relevant to the work of GIZ, but will also be useful to other organisations involved in professional cooperation management.

In other words, the time had come to update the Capacity WORKS manual. At the same time, other national and international organisations are becoming increasingly interested in this practical knowledge. Their interest is further boosted by the growing importance of cooperation in all areas of society. So GIZ is now making its expertise available to all interested readers. This version of Capacity WORKS will therefore not focus primarily on the specific context of international cooperation. We have set out to describe basic concepts and ideas in ways that are easy to understand so that they can be put to productive use in all kinds of cooperation systems, in all kinds of settings.

The model:
an overview of Capacity WORKS

Capacity WORKS is a model that enables users to successfully manage cooperation systems. It is based on various elements that are mutually complementary.

We will now briefly outline these elements.

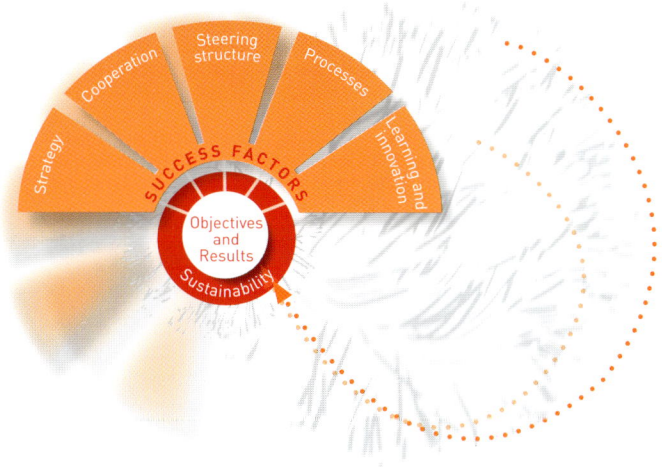

Figure 1: Capacity WORKS

Any cooperation system emerges and develops in order to achieve objectives and results that have been agreed between the actors involved. If they are to facilitate sustainable changes, the objectives and results of a cooperation system must be drawn from within the particular social context. To formulate objectives that are sustainable, we recommend striking a balance between social responsibility, ecological balance, political participation and economic capability.

This means combining the two core ideas outlined in the beginning, i.e. the guiding principle of sustainable development and the capacity development approach. These permeate all elements of the model, creating a focus on the willingness to change and the proactive management capacity of the actors involved. A process of negotiation between all the actors involved ensures that joint objectives are clearly formulated, attractive and realistic. The chapter on objectives and results explains this in detail.

These challenges are tackled using the five success factors, referred to as 'SFs' for short. These represent different perspectives to be adopted when systematically managing a cooperation system: strategy, cooperation, steering structure, processes, and learning & innovation.

The project is managed on the basis of these success factors. This also involves determining what contributions each of the individual cooperation partners will make.

To be successful, any project needs …

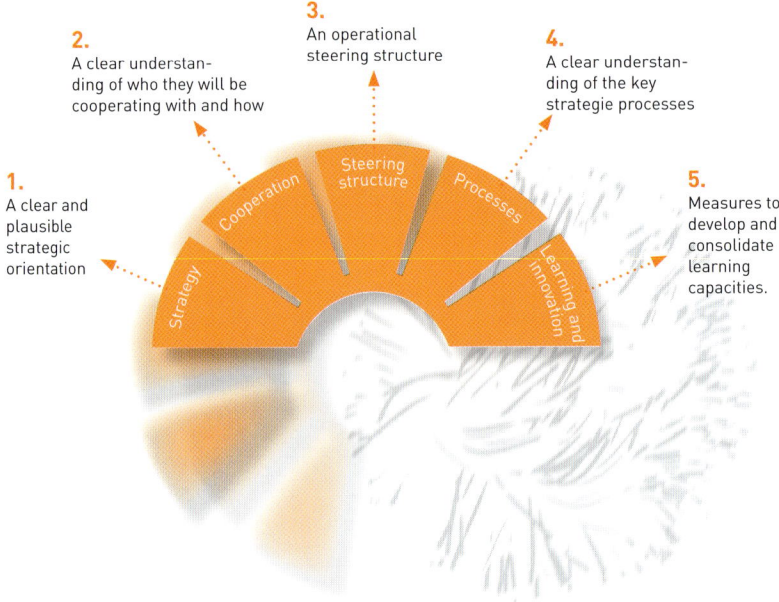

Figure 2: The five success factors

At the end of the present section each success factor is shown with its own motto, which high-lights its particular perspective. The mottos are supplemented with key questions that focus attention on specific aspects which have proved crucial in successful cooperation management. This provides the reader with a brief overview of the content of each success factor.

The introductory chapter is rounded off with some ideas on cooperation systems. In a next chapter the manual goes on to discuss objectives and results. This is followed by five chapters that describe each of the success factors in full detail. The description of the model is followed by the toolbox. This contains tools for each of the success factors that provide appropriate ways of reaching sound management decisions.

But what distinguishes the management of cooperation systems from management within organisations? The map of two logics graphic illustrates some key basic ideas for working with Capacity WORKS.

The map of two logics

In everyday life we use the term 'cooperation' all the time. This refers to the way in which different actors work with each other in order to produce results. Wherever cooperation takes place, it is also managed. Anyone with experience in dealing with organisations is familiar with the need for cooperation and management: teachers and school principals cooperate with each other, nurses and doctors do the same thing in hospitals (hopefully also across departments), production and marketing divisions discuss manufacturing operations, and ministerial policymakers work with the administration. At the same time, almost everyone has at some point experienced cooperation in organisations not working as well as it might.

Capacity WORKS is a model for the successful management of cooperation arrangements involving more than one organisation (inter-organisational cooperation systems). So, does Capacity WORKS also help manage cooperation within single organisations? At this point a word of caution is required. Organisations and inter-organisational cooperation systems follow very different logics. This means that the way they work cannot be explained and managed using a single model.

Capacity WORKS was developed in response to the following question: How can we help make cooperation between different organisations that are jointly seeking solutions to societal needs, problems or challenges a success? To answer this question, we need to take a closer look at the differences between working in the context of inter-organisational cooperation systems, and working in single organisations.

For single organisations there are already enough good management models around. These include the European Foundation for Quality Management (EFQM), Six Sigma and the Balanced Scorecard, to name but a few. However, these management models are not suited to the specific requirements of cooperation systems.

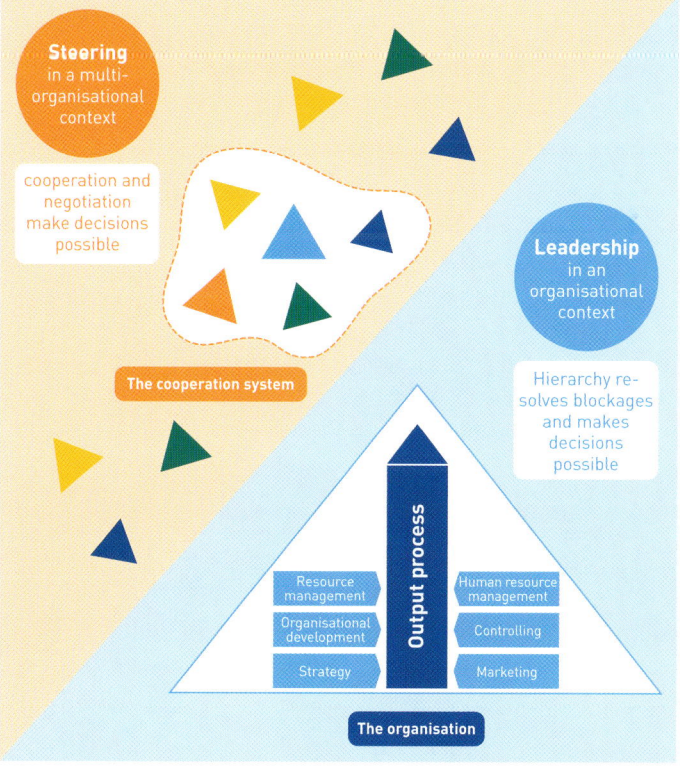

Figure 3: The map of two logics

The map of two logics explains why this is the case. It compares the different ways in which organisations and cooperation systems work, and provides the conceptual framework for understanding the context in which Capacity WORKS is applied. We will first of all look at the phenomenon of the organisation (bottom right of graphic).

The organisation

Organisations are social systems. This distinguishes them from other systems, such as technical systems or an ecosystem. One important feature of a system is that it requires boundaries in order to distinguish itself from its environment. These boundaries are used to define what belongs to the system and what does not. Social systems comprise at least two people. At the same time, social systems are able to relate to their environment. Organisations are a particular type of social system.

Defining objectives

Why do we need this kind of social system that we call an 'organisation'? Organisations are always responses to specific societal and individual needs. Organisations develop and specialise in order to deliver solutions to problems in a given society. For example, hospitals supply patient groups with medical services, public administration organisations deliver public goods, and commercial enterprises explore market needs which they then satisfy by supplying products and services.

Membership

To maintain their sustainability and ensure their survival, organisations clearly demarcate the boundaries that distinguish and separate them from their environment. Who is part of the organisation? Who is not? This question is answered using the criterion of membership. Usually a contract is drawn up that describes specific rules for entry and exit. Agreements are also often reached concerning the nature of remuneration, leave entitlements, the limited- or unlimited-term nature of the membership, and rewards and sanctions. It is important to highlight this, because members are not tied to an organisation 'body and soul' (i.e. constitutionally); they are bound only by their membership role. As well as being members of this organisation, people also operate in many other roles in their professional and private lives. This means they can also belong to one of several groups of stakeholders of the organisation.

Basic features of organisations

In the course of their history organisations develop and acquire a 'rationale of their own'. They strive to become 'immortal'; regardless of who their current members are, they form their very own 'DNA'. One basic element of organisations is decision-making. It is true that decisions are taken by human beings made of flesh and blood. However, once these decisions are established they develop a life of their own. This is very easy to spot in organisations that have already existed for many years. Members, including line managers, come and go. Yet the structures, processes, rules and rituals often remain in place for decades and change only slowly. This is due to the definitive or 'DNA-type' decisions that answer fundamental questions about how the organisation works. Why do we exist as an organisation? What are our tasks? How are we organised as an organisation? What are our expectations concerning how the members of the organisation behave?

These definitive, strategic decisions are reflected in all the structures, processes, rules and rituals that set the framework for everyday life in the organisation. They ensure that the organisation's basic way of working, the roles of its members and the expectations remain in place, even when there are changes in personnel. This often leads to an astonishing tendency toward inertia in organisations. In other words, organisations are more than just the sum of their members. Through its structures, processes, rules and rituals the organisation makes itself partially independent of individuals, thus ensuring the stability it needs in order to survive in the flux of change.

Decision-making in organisations: leadership

If we believe that organisations each have their own rationale, this has consequences for our understanding of management. On this understanding, 'line management' leadership is not a task performed by leaders who hold their positions because they possess a specific type of charismatic personality. It Leadership continuously supplies the organisation with the decisions it needs in order to ensure its own survival. This specialised function differs from the manifold technical tasks an organisation must perform in order to deliver its outputs.

In practice, this function of organisations – i.e. ensuring the ability of the organisation as a whole to survive – will be more or less well developed. Depending on the organisation it will be performed part of the time by designated line managers, and part of the time by other members of the organisation, sometimes within intelligent organisational structures and processes.

In modern societies line managers usually can no longer draw on traditional sources or ascriptions of authority such as background, education or power. Today's line managers must generate this authority anew every day through communication, in order to retain acceptance. This means they must always think carefully before invoking the power of hierarchy.

For the organisation as a whole, line management-based leadership is performed chiefly in six areas of activity, all of which are geared to disrupting the natural tendency of organisations toward inertia[1]:

1. Strategy development: orienting the organisation toward future trends

2. Human resources management: ensuring the workforce's ability and willingness to perform

3. Marketing: orienting the organisation toward the needs of its environment and the market

4. Resource management: securing the resources needed by the organisation to perform its tasks

5. Organisational development: finding the right organisational forms for generating demand-driven institutional performance

6. Monitoring: establishing appropriate self-monitoring mechanisms that allow key dimensions of the organisation's status to be measured swiftly and reliably.

The core task of this special function of line management is to continuously supply the organisation with viable decisions, and to resolve deadlock and conflicting aims within the organisation by communicating with its members.

Cooperation between several organisations

Capacity WORKS was developed for purposes of managing cooperation systems. This means we must examine the phenomenon of cooperation between several organisations (see top left half of graphic).

Since organisations often cannot meet the demands placed on them on their own, they must enter into cooperation arrangements with other actors. The organisations involved then face the challenge of 'getting into shape' so that they can operate successfully in these cooperation systems. This means they must develop the appropriate capacities. What is right for one organisation need not be suitable for all the other cooperation partners involved. Unlike in the context of a single organisation, decisions on joint objectives and the specific contributions to be provided by the parties involved are supplied not through line management leadership, but through processes of negotiation between several actors.

So what are the specific features of cooperation systems that distinguish them from the single organisation? Seen from the management perspective, where are the key differences that we need to be familiar with if we wish to operate successfully in cooperation systems?

Different ways of setting goals

Each organisation involved in the cooperation system will have its own goals and decision-making premises that shape its everyday activities. Very often these differ from the goals and decision-making premises of the other cooperation partners. The challenge is to negotiate a viable goal for the entire cooperation system. This presupposes that the cooperation partners recognise and acknowledge that they are dependent on each other. This dependency always arises in situations where a benefit is to be jointly generated that no single actor could achieve on their own.

In order to pursue joint objectives within a cooperation system, organisations partially waive their autonomy. The way in which an organisation works may perhaps not (yet) provide for decisions to be made in joint responsibility. That organisation may therefore be strongly tempted to transfer its own logic onto the cooperation system. A process of negotiation may also touch on sensitive areas of specific structures, processes, rules and rituals within the organisations involved. And in some cases the organisations themselves may have to change in order to operate effectively within a cooperation system.

Differences in terms of affiliation versus membership

A further key difference between cooperation systems and organisations involves the question of affiliation versus membership. In cooperation systems the forms and boundaries of affiliation are more flexible and more permeable through time than organisational membership. Cooperation is based on successful negotiation with the other cooperation partners, and is characterised by a high degree of voluntariness. If an actor calls into question the goal of the cooperation system, their participation in the system may then itself be called into question. Whether or not an actor will participate is a decision that always depends on a process of joint negotiation. Just as individuals are attached to their organisations only by virtue of their role and not 'body and soul', the cooperation partners and their organisations also retain some of their identity within the cooperation system. In fact they devote only some of their attention, some of their resources and some of their time to achieving the joint objective.

Different ways of reaching decisions: steering

The importance of decision-making in the context of line management-type leadership was outlined above. In cooperation systems, decisions also have to be reached in order to guide and coordinate the cooperation. How are these decisions reached? In these contexts, Capacity WORKS talks not about line management or leadership, but about steering.

In organisations, line management means that decisions can (if need be) be brought about through hierarchy, and deadlock thus resolved. In cooperation systems the option of using hierarchy in this way does not exist. In the course of time cooperation systems usually do form a steering structure that supplies the cooperation system with decisions – ideally in a way that is transparent for all parties involved. However, these decisions are generated through processes of negotiation that are more or less formally structured, depending on the cooperation system. Any attempt by a cooperation partner to bring about decisions by hierarchical behaviour is incompatible with the logic of a cooperation system, and threatens its existence. Here actors must avoid falling into the trap of assuming that the logic of their organisation is per se the best one.

This is important, because the cooperation partners remain autonomous in deciding whether and to what extent they wish to cooperate or not. Each cooperation partner makes their own contributions or inputs to the steering of the cooperation system, and is more or less effective in influencing it. These steering inputs involve actions or communication by actors, i. e. the performance of specific activities, or no action at all. Whether or not the cooperation system always absorbs these steering inputs in the way the actor providing them would like is something the actor cannot control.

When different partners in the cooperation system provide a large number of steering inputs, the process takes on a momentum of its own: the system begins to steer itself. This dynamic occurs regardless of whether it is conducive to achieving the goals of the cooperation system or not. It therefore makes sense to create steering processes that harmonise and coordinate these steering inputs.

Implications

The requirements created by line management differ from those created by steering. Almost always, representatives of organisations within cooperation systems operate on both sides of the map: in everyday practice, they often swap sides by the hour. As line managers, they may be involved in taking decisions on the contribution made by their own organisation. A few moments later they find themselves engaged in processes of negotiating the contributions of their own organisation with the other actors in the cooperation system.

Practice has shown that actors find it much easier when they are clearly aware of which context they are operating in at a given point. They then become more aware of the need to develop an appropriate inner attitude for each context. Anyone attempting to 'line manage' or 'lead' in a context of cooperation will be shown a red card by the cooperation partners involved, and rightly so.

Trying to apply the logic of steering in a context of line management leadership, however, is equally doomed to failure. One consequence of this can be organisational paralysis caused by an absence of managerial decision-making. When negotiation processes are created that cancel out established line management mechanisms, important decisions may be withheld from members of the organisation. These decisions, however, are necessary in order to resolve deadlock and conflict by means of hierarchy.

This is why Capacity WORKS focuses on how to successfully manage cooperation. The model supports users in identifying the right forms and content for negotiation processes in cooperation systems. We will now outline in just a few pages the key ideas contained in each of the five success factors (SFs). The conceptual thinking underlying them and the key questions provide rapid insight into the specific perspective on the management of cooperation systems contained in each of the success factors, and complete the model.

The success factors – an overview

The success factor 'strategy'

Motto: Negotiate and agree on the strategic orientation

According to one possible definition of strategy, good strategy is manifested as a 'pattern in the stream of decisions' (Henry Mintzberg). The strategic orientation of a cooperation system must match that of the organisations participating in it. This kind of pattern in the stream of decisions can only arise if and when the actors agree to negotiate one or several objectives with each other. This willingness has consequences, because in turn it also affects the strategies of the organisations involved.

Strategy development is a demanding task, because it requires the actors to develop a shared perspective. The key question is: Are we doing the right things? The actors are required to consider options which they perhaps initially find disagreeable. They must reach a joint decision that is both supported by the cooperation system, and supports it. In other words, the decision and the cooperation system support each other.

The process comprises various steps, all of which are equally important: (1) analyse (2) devise options (3) decide (4) elaborate the strategy (5) intregate into operations. If the actors omit one or several steps because they believe that sufficient clarity already exists, then they miss an important opportunity. What they miss is the opportunity to engage with each other. Although this may sometimes be difficult, it does allow the actors to deal with each other honestly and develop a joint perspective that is realistic. The SF strategy shapes the spaces for communication that allow this engagement to take place.

By engaging with each other and developing a joint strategic orientation the actors involved are able to clarify expectations within the cooperation system, and expectations of it. This will make clear which paths toward implementing the objectives and change will be pursued, and which have been discarded. The process of engagement motivates actors within the cooperation system to pursue the objectives with determination, and encourages the organisations involved to commit themselves. The joint strategy steers action toward areas of potential and energy for social change. It makes efficient use of existing resources and capacities within the cooperation system, and creates leeway for actors to act within the strategic framework.

It is helpful to ask the following **key questions** when developing the strategy:

- How does the sector or area of social concern 'work' at the moment?

- What strategies for change are being pursued by the actors operating in the sector?

- What joint objective can the cooperation partners agree on?

- What strategic options are available for achieving this objective?

- What strengths can be developed? What weaknesses should the strategy respond to? What opportunities and energy for change should be harnessed? What risks need to be taken into account in this context?

- How does the strategy respond to the way the sector works, for instance with regard to political feasibility?

- What criteria will the cooperation partners apply in order to select a strategic option?

- Are activities and outputs of the cooperation partners mutually harmonised and coordinated?

- How will the development of learning capacities be integrated into the strategy?

The success factor 'cooperation'

Motto: Connect people and organisations to facilitate change

When actors decide to enter into relationships of cooperation with other actors this does not change anything fundamental as regards each of them acting according to their own will. Nonetheless, to a certain degree they do voluntarily restrict their own autonomy. When actors act as partners in a cooperation system they do not lose their identity, but continue performing their own tasks as an organisation. They simply need to divide up their energy accordingly. The energy that each actor must devote to cooperation is like a fuel that is both scarce, and expensive. Professional cooperation management helps build forms of cooperation that deliver results, while striking a balance between demands in the context of the organisation and demands in the context of the cooperation system.

The SF cooperation focuses inter alia on the actors involved or yet to be involved. Interests and attitudes toward change objectives are reflected on, as are influence and responsibilities within the area of social concern. Cooperative and conflictual relationships are analysed in detail, as are the roles of the actors involved and the appropriate forms of cooperation. The boundaries of the cooperation system are defined, which then determines which actors will assume joint responsibility in order to achieve the desired changes.

Networks are not systems of cooperation, as they perform highly particular functions and therefore also follow different rules. They do not possess the structures of a cooperation system, and involve cooperation that is considerably less binding. The distinction between cooperation systems and networks has far-reaching consequences for successful cooperation management. Depending on the objectives of the cooperation, the actors involved will select an appropriate form of cooperation.

It is helpful to ask the following **key questions** when establishing cooperation relationships:

- Which actors are relevant in the sector or area of social concern?

- What mandates, roles and interests do these actors have? How do they operate within the sector?

- What lines of conflict exist, and how can we deal with asymmetries of power within the cooperation system?

- Which actors must be involved in order to achieve the agreed objective? Whose participation is not necessary?

- What forms of cooperation are appropriate?

- Do the various actors possess the resources necessary to achieve the agreed objective?

- What strategically important resources outside of the sector (local, national or international) would it also be worthwhile for the project to acquire? Which individuals, organisations and networks outside of the cooperation system might be considered as external partners for the project?

- What comparative advantages make the cooperation system an attractive partner for complementary cooperation?

The success factor 'steering structure'

Motto: Negotiate the optimal structure

Like organisations, cooperation systems must also be supplied with decisions. In cooperation systems, decisions are always taken in processes of negotiation between the cooperation partners. The steering structure provides social spaces for these processes of negotiation. The option of applying the principle of hierarchy is not available. Each of the actors involved attempts to provide inputs to the steering process, in the hope that these will be accepted by the cooperation system as a whole.

The steering structure provides the cooperation system with: strategic and operational decisions, conflict management, resource management, operational planning and monitoring of implementation. In particular, the steering structure delineates the rules, roles, mandates and responsibilities in the decision-making processes. It is helpful to distinguish between politico-normative, strategic and operational levels of steering. This enables steering tasks to be delegated, and for instance relieves high-ranking decision-makers of having to take decisions that can be taken by people at the next level down who are better informed. Thus, applying the principle of subsidiarity creates greater overall acceptance of the steering structure among the actors involved. Even projects of limited duration that are designed to facilitate social change in an already existing and permanent cooperation system also need a steering structure. Wherever possible, these should be tied to those steering structures that already exist. Many demands are placed on the steering structure of cooperation systems. Ultimately, though, they are judged by only two criteria: The optimal steering structure must be functional with respect to the targeted objectives and results, and must be appropriate for the complexity and the scope of the task in hand. The more complex the objectives and tasks of a cooperation system are, the more sophisticated and complex the steering structure will usually have to be.

It is helpful to consider the following **key questions** when negotiating a steering structure:

- How are decisions reached in the sector or area of social concern?

- What do we believe the steering requirement in the cooperation system to be? Does the co-operation system require additional steering structures, or can it use structures that already exist in the sector?

- How will the steering structure cope with the diversity and scope of the tasks to be steered, and the associated risks?

- How will broad political backing be created for the objectives and the change process?

- What measurable variables will steering decisions be based on? What kind of monitoring system is required in order to support steering of the cooperation system?

- How will decisions concerning resources be negotiated, agreed and implemented within the steering structure?

- What does the plan of operations for implementing the strategy look like?

- How can the project steering structure be designed so that a model emerges which fosters the culture of cooperation in the cooperation system in the long term?

The success factor 'processes'

Motto: Design processes for social innovation

Social innovation emerges from a process of societal change that is rarely linear, and usually cannot be planned. Nonetheless, the cooperation partners within a cooperation system do decide to attempt to drive innovation in a structured way.

First of all they analyse the processes in place for delivering services that are relevant to society. The actors then define the points at which change is supposed to take place. Following that, change processes are initiated that mainstream innovations in the routine operation of the cooperation system. This requires the establishment of close links between the permanent cooperation system and the limited-term project designed to facilitate change.

The SF processes focuses on both aspects: First of all the processes within the area of social concern to which the change processes relate are analysed. Secondly the internal processes in the project that aim to prompt these changes in the sector are established and reviewed. These internal processes relate to cooperation among all the actors involved in the project.

One of the key elements of the SF processes is the so-called process map, which provides a visual overview of a cooperation system. Based on the outputs that various actors generate together, the processes are categorised according to different process types. These distinctions are then used to analyse the status quo of a cooperation system and determine the need for change. The output processes are the processes that relate directly to the objectives of the cooperation system. The cooperation processes underpin the output processes by coordinating the various actors. The learning processes are necessary, because this involves the actors appraising the quality of service delivery in the sector and making needed changes. The support processes are packages of tasks that underpin all the other types of process. The steering processes are the ones that set the legal, political and strategic framework for the other types of process

While the process map provides a strategic view of the sector, the process hierarchy supports operational planning as well as in-depth analysis. This is used to visualise selective processes in further detail by depicting their sub-processes. The degree of detail needed will always depend on the requirements of the specific case.

It is helpful to us the following **key questions** when focusing on processes:

- What are the relevant processes in the area of social concern, and what form do they take?

- How do the core processes (output, cooperation and learning processes), steering processes and support processes interact? Where do strengths and weaknesses exist?

- In which processes in the sector should the project invest in order to gain leverage?

- Through which processes will the project influence the management of processes in the sector?

- What will the project's output processes need to look like in order to achieve this?

- To what extent can the change processes be transferred so as to support social innovation in the cooperation system? Do the change processes serve as models for creating social innovation in the cooperation system (and beyond)?

The success factor 'learning and innovation'

Motto: Focus on learning capacity

Successful cooperation management focuses our attention on the fact that while pursuing capacity development, learning capacity must be strengthened on all levels. Within the society, frameworks are adjusted and cooperation relationships improved. Organisations learn to help achieve the joint objective while continuously raising quality and facilitating further learning. People in organisations develop their competencies and shape learning processes in ways that can help generate sustainable results in their respective settings. All this puts conditions in place for launching and realising innovation.

We can explain how organisations and cooperation systems learn, and how innovations become established, with reference to the three basic mechanisms of evolutionary theory. In organisations and in cooperation systems, minor or major variations from the established routine emerge at various points either as a result of planning or spontaneously (variation). If too many variations occur, their sheer diversity can leave the members of organisations and cooperation system partners uncertain as to how they should act. This creates a need for decisions, based on line management or steering, to select from among the available variations the ones that are best suited (selection). After selection, measures are required to stabilise the innovation within the system (stabilisation or re-stabilisation). This means that rules, structures, processes and rituals are reviewed, and where necessary adjusted: new routines emerge. The cooperation system or the organisation then gains the stability it needs in order to survive.

Learning and innovation are often generated by individuals who see new opportunities and potential, or quite simply a mismatch between the way things are supposed to be and the way they are. Competency development places the human being at the centre of her potentiality and is an integral component of the capacity development approach. Organisational development activities become more effective and sustainable when they are supported by the development of competencies and learning networks at the level of individuals. As change agents, individuals can make processes of exchange more efficient, initiate new orientations in their own settings, and consolidate learning by acting as disseminators. At the same time, capacity development processes at the level of organisations and societies create an enabling environment for effective and sustainable competency development.

It is helpful to consider the following **key questions** when developing learning capacities:

- What innovations should be mainstreamed (scaled up) in the area of social concern?

- What learning goals do the project objectives implicitly contain?

- What are the learning needs on the three levels of capacity development?

- What capacities are present within the cooperation system for developing strategies, making cooperation sustainable, taking decisions and managing processes? What action is needed as a result?

- What measures will be taken to ensure that specific project actions lead to learning? Do the lines of action match and reinforce each other? What additional interventions also need to be initiated with regard to the learning needs?

- Bearing in mind the mechanisms of variation, selection and (re-)stabilisation, how will the project support learning and the mainstreaming of learning processes within the cooperation system?

- How will lessons learned in the project be analysed and documented so as to support the development of learning capacities within the cooperation system?

Cooperation systems – permanent and temporary

When we speak of cooperation systems we need to carefully distinguish between two different types. On the one hand there are permanent cooperation systems that have emerged to manage a certain set of societal problems or achieve a joint goal. This type of cooperation system aims to supply the society with specific services, such as water or health care. These are permanent cooperation systems that deliver public goods or services for an area of social concern.

By contrast, there exist temporary cooperation systems that supply policy fields with social innovations. As we mentioned at the beginning, these are referred to as 'projects'. Project objectives are always clearly oriented toward a permanent cooperation system in which changes are to be mainstreamed. Projects are spaces where these changes are piloted before being scaled up across the sector. In other words, new or different processes are established in order to deliver services in a relevant social context. This can be manifested for instance in changed forms of cooperation between the actors involved, in proposals for the amendment of legal and policy frameworks, or in initiatives for change in relevant organisations, in response to new demands.

One key conceptual challenge is to recognise the fact that a project is not a laboratory that can be used like a sterile area in which to develop prototypical solutions for the challenges of real life. A project must not be cut off from or treated independently of the permanent cooperation system. The actors involved must not put aside the concerns of the permanent cooperation system when they enter the laboratory; if they did this would mean they were forgetting about their normal day-to-day business when cooperating in the project. It would be more appropriate to compare implementing a project with repairing a vehicle with its engine running. The temporary cooperation system must be designed so that it relates to the demands and possibilities of the permanent cooperation system, and generates a perceptible benefit for the actors. To achieve this, an appropriate balance needs to be struck between routine operations and stimulus for change.

To make this possible, a project must always be conceived from the perspective of the relevant social context. What does this mean? This becomes clear when we consider the following example. A project is designed to improve the access to financial services for small and medium-sized enterprises (SMEs). This means that the project objective relates only to a segment of the financial sector as a whole. Nevertheless, the project must fit into the logic of the sector, for instance with regard to the role of private banks. It may be that certain actors will first of all have to be persuaded that the project objective is worth pursuing.

At this point we need to state one thing very clearly: the project will not replace the sector. Nor does the project exist in addition to the existing structures, and nor will it hand over turnkey solutions at the end. It is important to avoid creating parallel structures and duplicating work, so that sustainable solutions can be developed that can be permanently incorporated into the logic of the sector. A temporary cooperation system is carried by the engagement of various actors and is a platform for all those involved to achieve the joint objective. Of course the individual organisations will derive particular benefits of their own from the cooperation as they make their own specific contribution, but the focus will be on developing sustainable solutions within the cooperation system.

The capacity development trilogy

So how can changes be initiated sustainably in a given social context? The graphic of the capacity development trilogy shows how the permanent cooperation system, the temporary cooperation system and the specific contributions made by the cooperation partners fit together and interact.

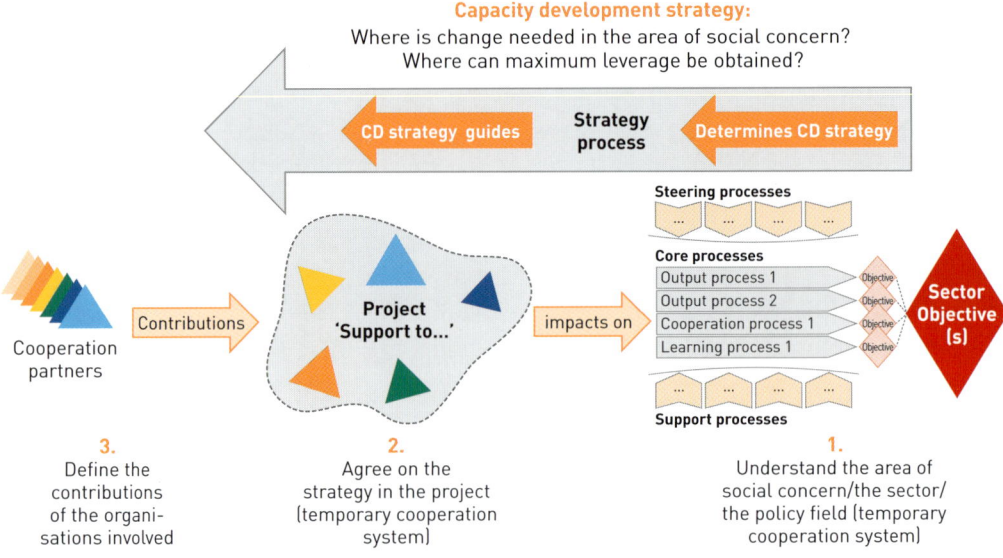

Figure 4: The capacity development trilogy

Starting on the right we can distinguish three elements: A sustainable change is to be facilitated within a sector. This means we have to establish a basic understanding of how this permanent cooperation system works. This means assessing what capacities of people, organisations and the society are required for the desired change. The ultimate aim is to develop the capacities of the people, organisations and society involved to manage their own sustainable development process and adapt to changing conditions.

The temporary cooperation system (the project) in the middle of the graphic is designed such that it corresponds to the logic and the potential for change of the policy field, and is responsive to the understanding of the permanent cooperation system. Who needs to change what, so that the desired objectives and results will be achieved? How can that take place? Who needs to learn what on which level, so that the changes can be sustainable and mainstreamed?

The project uses the contributions made by the cooperation partners involved (left side of graphic). These may be of very different kinds, e. g. financial inputs, work, services or knowledge. The project must make intelligent use of the existing resources made available by the cooperation partners from their respective organisational contexts.

Where exactly does Capacity WORKS support this? The management model Capacity WORKS enables the actors involved to monitor and analyse the setting of a project – i.e. the permanent cooperation system in which changes to take place – and draw conclusions for the change strategy. Capacity WORKS also helps design and implement the project. Objectives are jointly defined, possible cooperation partners are selected, a bespoke capacity development strategy is developed, decision-making mechanisms are established and specific activities are implemented at the operational level.

The key element for achieving the desired sustainability of the changes is the logic of continuously establishing a conscious link between the policy field and the project. The permanent cooperation system must incorporate these changes and integrate them into its routines. To achieve this it requires inputs on various levels, all of which relate to the joint objective: at the level of society with its frameworks and established relations of cooperation, at the level of specific organisations, and at the level of human individuals who in their own way help ensure that the sector works.

Capacity development for renewable energy use

If renewable energy use is to be developed in a country, one option would to implement a project to support the installation of solar panels in private households. Among other things, this would require legal provisions to regulate access to state funding for research projects. This could be used to drive the development of low-cost technical solutions.

Public-sector actors, universities and business associations are already implementing joint research projects. These actors are working jointly to achieve the objectives of the new project, and by doing so are initiating a temporary cooperation system. To enable them to make an effective contribution toward the project, individual organisations need to be strengthened, for instance regarding their capacity to form project teams at short notice, coordinate the work of these teams across internal departments, and integrate the results of the work into their organisational processes. The individuals working in the organisations involved also need to develop new competencies so that the cooperation system can generate the anticipated result. If they are members of a project team they may require training in the fundamentals of project management. If they are line managers, they may need to develop competencies for leading the organisation successfully through this change process. If they are trainers for the new technical solutions, the individuals concerned must know where they themselves can learn more about these technologies, which teaching methods they should use to best transfer their knowledge, and how they can initiate learning networks.

When the various levels of capacity development work together in concert, the temporary cooperation system will transfer this innovation into the permanent cooperation system. At the same time we may also assume that the actors involved will succeed in managing new tasks in the future, as they have now developed their competency for change as a whole.

In other words, successful capacity development is the key to sustainable development. Both pre-suppose that the actors involved are willing to change. If a project does not succeed in tapping into and using this willingness to change, it is highly likely to fail.

The management model Capacity WORKS support users in visualising these constellations itera-tively and from different perspectives. This facilitates objectives-oriented communication between the actors involved. By engaging with each other in this way they establish a shared picture of the reality. In this way the actors find the spaces they need to design a path for change that will motivate all those involved to make their contribution.

Using Capacity WORKS

How useful the oracles were in ancient times. A clear question, one person with the ability to read the auspices – sometimes with and sometimes without a sacrificial offering – and right away those seeking advice would have a basis on which to make their decisions. The management model Ca-pacity WORKS differs from the oracles in many respects, even though there are similarities at the beginning and the end of the process. One similarity at the beginning is: the clearer the question to be addressed, the easier it is to find the right way of addressing it. And at the end, like the oracle Capacity WORKS delivers the basis on which to take decisions.

Between the beginning and the end, though, everything is different. At least in the case of coop-eration systems, the answers to questions usually cannot be provided by people on the outside. We might mention in passing that in ancient times too, the oracles only rarely delivered answers that were self-explanatory. Croesus, the immensely wealthy king of Lydia in 546 BCE, should for instance not have been so quick to interpret the following oracle of Delphi: 'If Croesus crosses the River Halys, a mighty kingdom will fall'. Brimming with confidence, Croesus then took on King Cyrus II of Persia. Yet it was not the latter's kingdom that fell, but Croesus' own.

Capacity WORKS is built on the principle that the signs of the times will be read by the cooper-ation system actors themselves. The actors themselves must set out in search of suitable answers, formulate their assumptions, and review, confirm or discard them. Responsibility for taking the right decisions always rests with the cooperation partners themselves. Rarely is it possible to say with absolute certainty what is right and what is wrong, because cooperation systems are social systems that operate in complex societal settings. All of this is inherently unpredictable. So, Ca-pacity WORKS is not an oracle!

When the management model was first introduced at GIZ, many people were surprised by the fact that when addressing questions in cooperation systems it was considered beneficial to raise further questions. This seeming paradox is based on the assumption that no one is in a better po-sition to decide on how to manage cooperation systems than the actors themselves. The questions that Capacity WORKS supplies ensure that the actors use their own implicit knowledge. External expertise can be helpful, but does not deliver sustainable solutions. These can only be developed by the actors themselves developing a shared perspective on their own reality. However, this also means to some extent abandoning the predictability of outcomes.

'Everything is possible, but nothing is certain!' Written in graffiti on the steps of Delphi this would have been entirely out of place, yet it is extraordinarily accurate in capturing the attitude to life found in modern societies. It is interesting to imagine how this piece of graffiti would be read by a trade union official who has just informed a group of factory workers that short-time working hours were being introduced. Or to imagine what the head of department at the ministry of labour working on a decision-making proposal for social reform would make of it. And what would the engineer think who was just about to successfully complete a process of research into new production methods? Does this message arouse a longing for certainty and predictability? The acceptance of constant change? A willingness to explore newfound spaces of agency?

The management model Capacity WORKS helps users discover new spaces of agency for social change processes, and is designed for all those who wish to better understand or manage coop eration systems.

Fitting the various elements of Capacity WORKS together

Depending on the desired depth of results, in many cases it will be sufficient to structure a process of reflection simply by applying the logic of the five success factors using the appropriate key questions. If a topic needs to be addressed in greater detail, we recommend dipping into the toolbox. Please remember: Never use a tool without first of all having a question to begin with. Many tools will help users jointly approach a variety of topics through analysis and reflection. Some users may hope that by simply 'filling in' the tools the solution to the question (e.g. regarding strategy) will come out at the other end as if it were coming out of a funnel. Unfortunately that is not the case. As in life itself, as well as analysing and reflecting upon the situation, the actors must also have the courage to steer the discussion toward a decision.

Selecting the 'right' success factor or the 'right' tool, at least to begin with, is always a matter of knowing what question they will be used to address. The five success factors are closely interlinked and cross-referenced, and offer five different perspectives on the reality of the cooperation system. Like searchlights, the success factors illuminate things from different angles. Any areas left in the shade are then brought to light the next time round by swivelling the searchlights in new directions. Usually, applying one success factor will inevitably open doors to the other ones. Essentially this means that there is no 'right' or 'wrong'. The most important thing is what works and what does not.

Here are some suggestions that illustrate how the tools might be employed in very different ways:

- A tool can be used by an individual or a small group from within the cooperation system to prompt ideas regarding a specific question. Elements are identified and used as a basis for a discussion. This is certainly the shortest and most time-saving way to use the toolbox.

- A group of actors in the cooperation system require a basis on which to discuss a specific question. Here, the time required to use a tool will be dependent on the heterogeneity and the size of the group. Depending on the complexity of the task in hand, using the tool in this way can last anything between half a day and a workshop of around a day and a half.

- Using the tools will take up the most time when the actors in the cooperation system use the toolbox to take fundamental decisions, such as the decision on their project strategy. In this case it may be necessary to hold workshops lasting several days.

The same applies when using all other elements of the management model. The key questions for the success factors can be discussed systematically within the cooperation system in order to obtain an overview of the joint orientation: They can, however, also be used by individual actors to prepare for a meeting or a workshop. The success factors supply comprehensive expertise on issues that arise in cooperation systems, providing guidance for managing those systems. At the same time the headings – strategy, cooperation, steering structure, processes, learning & innovation – provide all users with a clear idea of how they can be used in processes to negotiate cooperation.

Using Capacity WORKS playfully

Experience with Capacity WORKS has shown that the more playfully we use the management model, the easier things become. When we say 'playful' we do not mean the opposite of 'serious'; we mean needs-based, flexible and willing to try out the model and the tools. During childhood this was the way most of us naturally went about doing things. Playing games often involves agreeing on certain rules that serve as instructions on how to play. However, once these instructions prove cumbersome or do not work very well (the participants themselves normally decide when this is the case) the instructions are either thrown away, modified or rewritten.

A huge amount is gained by adopting this open mindset when using Capacity WORKS to manage cooperation systems. For instance, a particular tool might not fit the question perfectly down to every last detail. In that case it should be supplemented, or only parts of it used, or it should be abandoned altogether. Nor does 'playful' mean 'arbitrary', because the context in which Capacity WORKS is used is clearly the cooperation system (whether permanent or temporary).

How to proceed with Capacity WORKS

Capacity WORKS offers various angles from which to take a structured look at cooperation systems. This helps users to assess the status quo of an area of social concern, and on that basis identify realistic objectives and results for a project. The model is also used later on when the project is being managed and the strategy implemented.

In other words, Capacity WORKS keeps one eye on the area of social concern (the permanent cooperation system) and the other on the project (the temporary cooperation system). It combines the two in practical ways in order to prevent two risks: action for action's sake, which is the result of inadequate analysis, and paralysis, which is the result of excessive analysis, which saps people's courage and leaves them with no energy for actual activities.

The management model also plays an important role in monitoring results achieved. It supports monitoring in the project, reviews it, and helps users check whether a cooperation system is moving in the desired direction. So how can all these functions of the management model be used?

The systemic loop shown in the graphic explains how:

In cases where a project is to be initiated in order to facilitate specific changes in a permanent cooperation system, a cyclical approach has proved beneficial. The first step involves gathering information. This is used to enable the participants to develop a shared and true picture of the permanent cooperation system. The five success factors provide helpful angles from which to approach this.

On the basis of the information collected, assumptions or 'hypotheses' can be formulated. 'Hypotheses' are assumptions and impressions formulated in positive

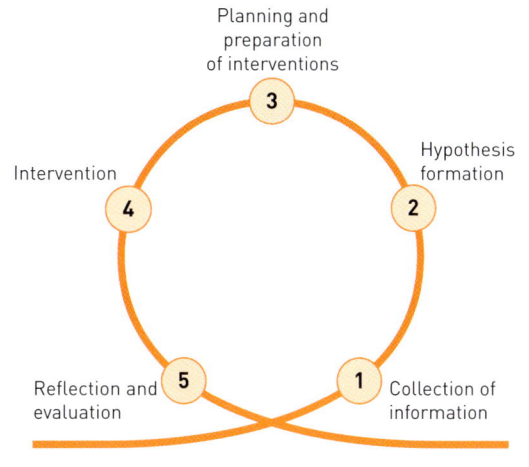

Figure 5: The systemic loop

terms on the basis of the information, data and observations collected. These represent an alternative to a specific 'solution', which of course can always be seen as 'right' or 'wrong'. Hypotheses are supplied as descriptions of states that over time can be either confirmed or refuted. A hypothesis that proves 'wrong' also always helps increase the amount of information available, because it rules out certain options. It can be helpful to ask the following questions when formulating hypotheses: How does the societal setting in which the project aims to generate results 'work'? What strengths and weaknesses, as well as opportunities and risks, can be identified? Against this backdrop, what objectives and results could realistically be achieved?

Building on that, joint objectives are negotiated and fundamental decisions taken that are important in defining the project approach and planning specific actions. What should the path to the desired change look like? Who will cooperate with whom, and how? How will decisions be taken? Which processes do we wish to influence, and what processes need to be established in order to do so? Who needs to learn what in order to establish the desired change successfully?

As the agreed measures are being implemented, the changes achieved are monitored and the entire 'architecture' of the project is continuously reviewed. A monitoring system, i.e. the systematic gathering of information on the progress made toward achieving the objectives, delivers the basis on which decisions can be taken for steering the cooperation system. This information should also be used periodically to test the hypotheses formulated at the outset. These hypotheses did after all play a key role in the project architecture. If the ambient conditions have changed, this will probably lead to an adjustment of both the project objectives and results, and its architecture. As the loop shows, this procedure does not end until the project has been completed and integrated into the routines of the permanent cooperation system. Once Capacity WORKS has done its job, the cycle begins again.

So how can practitioners use the manual in the everyday context of cooperation systems? The following example shows how:

1. Formulate the question and agree to address it with the relevant actors in the cooperation system.

2. Read the brief descriptions of the success factors together with the key questions, and decide which of the five success factors the question is most closely related to.

3. Read the full description of the success factor and form hypotheses on how best to address the question.

4. Browse through the toolbox for the success factor. Identify suitable tools, and if necessary make any adjustments.

5. Prepare an appropriate setting in which to address the question, and make the appropriate arrangements with the relevant actors.

6. Address the question together with the actors involved, and document decisions and results.

7. Agree further steps, including follow-up and communication of the results within the cooperation system.

Good practices

Working with Capacity WORKS at GIZ led to a vigorous debate on how GIZ supports sustainable change processes. The following principles have proved invaluable in our many years of consulting work to support cooperation system management:

- Projects should always be designed to **tap into existing energy for change**. Successful change processes are based on the will of actors within a cooperation system to initiate those changes. This will succeed where existing initiatives are harnessed and incorporated into the architecture of the cooperation system.

- The distinguishing feature of projects is the way they act as **catalysts**. They provide a platform on which the actors involved get together in search of solutions to issues of concern to a society. They provide spaces in which new forms of cooperation can be rehearsed before being integrated into the area of concern. Successful projects can serve as models for scaling up.

- Nothing motivates people more than rapid success. Without losing sight of the sustainability of results, it is a good idea to also focus on **quick wins** in order to motivate the actors concerned and boost their willingness to change. Positive experiences and joint success encourage people to place trust in their own ability to innovate. This lays the foundation for more comprehensive change.

- For projects to develop the necessary 'appeal' these success stories must be relevant to the social context. Users need to identify focuses of potential for change which thanks to their **leverage** will entail other changes within the sub-system.

- Usually, leverage is only achieved when capacities are developed on all three levels – i. e. **society, the organisation and the individual** – thus increasing proactive management capacity.

- Social change processes set in at different levels of society. Impetus for change must be generated as part of a **multi-level approach** encompassing the macro, meso and micro levels.

- Social contexts are always unique. This is why blueprints for change processes cannot work. The key to success lies in an **appropriate mix of methods** for change that is tailored to the specific cultural and political features of the system concerned.

- Changes within social systems are always complex and require professional inputs from different areas of specialisation. **Interdisciplinary approaches** can be oriented toward the needs of the actors.

- **Specific technical advisory services** will only succeed when combined with **policy and management advice**.

- Social change processes cannot be fully planned or steered. It is therefore helpful to develop **results hypotheses**, and continually review and test them under practical conditions. Developing visions that reflect a joint perspective of the different cooperation partners creates new spaces of agency for successful change processes.

These principles enjoy high priority in the design and implementation of projects in which GIZ is involved. The concepts and tools contained in Capacity WORKS incorporate these principles, and in so doing ensure that they play a role in the management of projects. The aim is always to harness the knowledge of the actors involved, generate fresh insights from the joint processes of reflection, and reach decisions which, if not necessarily the right ones beyond all shadow of a doubt, are nevertheless at the very least logical.

Objectives and results

Did you know that Aristotle philosophised about the nature of change processes? He drew a number of distinctions that are helpful in the context of modern cooperation systems. Aristotle's first assumption was that we can only understand change processes if we know what their causes are. We can equally well refer to changes as 'results'. In other words we are talking about the cause-and-effect relationships that lead to these results.

Changes or results can be planned or unplanned. When people plan them, they are pursued in order to achieve a predetermined objective. If they occur without having been planned, those on the receiving end will consider themselves either 'lucky' or 'unlucky'. Either they will be grateful to fate, or ask themselves whether they really could have done anything to avoid its ruthlessness.

An objective describes a positive state at the end of the planned process. The steps along the way help achieve that objective. Human behaviour makes sense when it can be linked to a particular objective or purpose. Objectives can be either more or less explicit.

When different actors come together in a cooperation system, the objectives should be as explicit as possible. Since the objectives guide the behaviour of the actors it is important to avoid major differences in interpretation. Ultimately the joint objective describes a positive state that the actors hope to achieve in the future. If this were to be interpreted in very different ways this might call the cooperation system itself into question. Aristotle himself noted that 'A conviction shared by all people has reality'.

Objectives are shared visions of the future

In other words, objectives are a shared picture of the future that represents a change in the status quo. In order to sustainably strengthen the proactive management capacity of the cooperation partners, it is absolutely essential that the actors involved participate actively. Yet what path should the actors follow in order to reach this future they hope for? Which cause-and-effect relationships should they base their actions on? What is the basis of their expectation that together they can create something new from the existing status quo?

The objective that the cooperation partners agree on gives meaning and direction to the entire change process. It is also closely intertwined with various aspects of the path for change:

- Changes always relate to a baseline. The actors' shared picture of the initial status quo sets the framework for the objectives that they consider possible.

- The actors in a cooperation system describe the path from this baseline to the objective by developing a strategy for change.

- The strategy requires them to describe the path for change. The success factors cooperation, steering structure, processes, and learning & innovation help the actors negotiate and describe with whom and how the objective is to be achieved, and through which specific activities.

- When implemented, these specific activities affect the status quo, thus bringing about change. The strategy sets a framework for this and provides key guidance. The progress of the intended changes should be continuously reviewed in order to supply the cooperation system with relevant information for steering.

This steering information will also be used to initiate an accompanying learning process to ensure the sustainability of the changes. This means that analysing the initial status quo is key to defining objectives for a change strategy. Unless objectives are formulated, no strategy can gain a clear profile. Without a clear strategy, actions can all too easily lead to activity for activity's sake.

Joint objectives strengthen cooperation

One thing becomes clear: defining objectives is as important as it is difficult. The more actors that are involved, the more complex the task becomes. The objective should be clear, so that everyone involved shares the same understanding of it. It should be attractive, so that the actors feel strongly committed to their objective. And at the same time it should be realistic, because hardly anything motivates people as much as success.

Meeting these demands requires a sound process of negotiation. The objectives should articulate changes that are both desired and feasible. This makes it necessary to clarify among other things what interests the actors are pursuing when they come out in favour of or against certain changes. Do these actors possess the necessary proactive management capacity, i.e. the power, interest, knowledge and resources needed to bring about these changes?

In other words, the key is to analyse the status quo. The actors jointly analyse the area of social concern in which the results are to be achieved. Invariably they then develop – usually on their own at the beginning – hypotheses concerning the cause-and-effect relationships in the cooperation system.

Since these hypotheses are often not made explicit or shared, the process of negotiation needs to prompt precisely this. Once joint objectives have been defined, the hypotheses help those involved to develop a strategy for the intended changes. The next step is implementation. Agreed actions by the cooperation system stimulate developments in the area of social concern that hopefully will lead to the expected changes. For this to succeed the actors involved must respond to the stimulus and use the results of the implemented measures. Proposals for mainstreaming a particular innovation are only meaningful if and when the cooperation partners see them as potentially useful and try them out. If an experience of this kind is considered a success, then the innovation can be accepted and thus help achieve the objectives.

A project to raise the quality of service delivery and efficiency of local governments

In pilot municipalities the project attempts to establish city offices that provide all services for citizens from a single contact point. This is based on certain hypotheses:

1. The quality of service delivery will increase when customers only have to approach a single contact point to obtain certain services, and no longer have to go from office to office.

2. Administrative processes will become more efficient when they are separated from face-to-face customer contact.

This means that the service point staff may have to learn many new things if they have only been dealing with some of the administrative tasks prior to that. At the same time the administrative processes 'behind the scenes' will need to be reorganised. The people involved will only support these changes if the objectives of service-orientation and increased efficiency make sense to them. The interests of those affected play an important role in determining whether or not changes succeed. The process can be supported by facilitating an explicit discussion of the hypotheses identified. If the results are rated positively in the pilot municipalities, this will provide a sound platform for transferring the innovation 'city offices' to other municipalities. In other words, carefully taking the various interests into account plays a key role in achieving the intended objectives and results.

So, the more explicitly the hypotheses on cause-and-effect relationships are worded, the more realistic the objectives will be. The more actors that share this vision of reality, the more likely this is to have a positive effect on cooperation. Consequently, objectives ...

- provide clear guidance for the joint activities of the cooperation partners and for their specific inputs;

- enable the available resources to be used to achieve the intended results;

- allow the planned path for change to be continuously reviewed.

If it turns out along the way to the objective that the intended results are not being achieved, then it is time to also closely scrutinise the objectives, and if necessary adjust them.

The results model – a shared vision of change

Cooperation systems are social systems. This means there are always complex and their responses are difficult to predict. Consequently, change can only be planned to a certain extent. Working with hypotheses helps deal with the uncertainty that this creates. The assumptions concerning which inputs will lead to which changes make dealing with the reality less complex. This creates new spaces for decision-making and agency for the path to change in the area of social concern.

The power of images can be used to document the joint understanding of this path to change in a way that all the actors can relate to. One way of dealing with this highly complex challenge is to construct models.

A model is required that visualises how the actors understand the cause-and-effect relationships involved in the intended change. This model should help to negotiate realistic objectives, and help the cooperation system to reach a number of fundamental management decisions when it begins its work.

The example below could come from any one of many regions in the world. Perhaps the actors will be supported by an international cooperation organisation, perhaps they will receive the support they need from a ministry or perhaps they will fund themselves.

The results model for regional tourism development

The initial situation is difficult, because the region has been economically dependent on forestry for generations. Within a few years several large companies closed down, including sawmills, companies in the furniture industry and haulage contractors. Unemployment soars. The region plunges into a kind of collective state of shock-induced paralysis.

A small group of actors from the local chambers of crafts and tourism seize the initiative. They invite the six local governments to a roundtable in order to explore fresh potential for development in the tourism sector. Due to the large number of undeveloped sites of natural beauty, it also proves easy to persuade the provincial government to develop a regional master plan for tourism. A project is then planned to operationalise this master plan and provide fresh impetus for tourism development. The journey begins …

The results model shown here makes explicit the key results hypotheses underlying the strategy for change that the participating actors wish to jointly implement. First of all they agree on the objective: A more conducive environment for tourism development is to be created by implementing a regional master plan for tourism.

After having analysed their initial situation in detail, the cooperation partners define the following areas as high-priority lines of action:

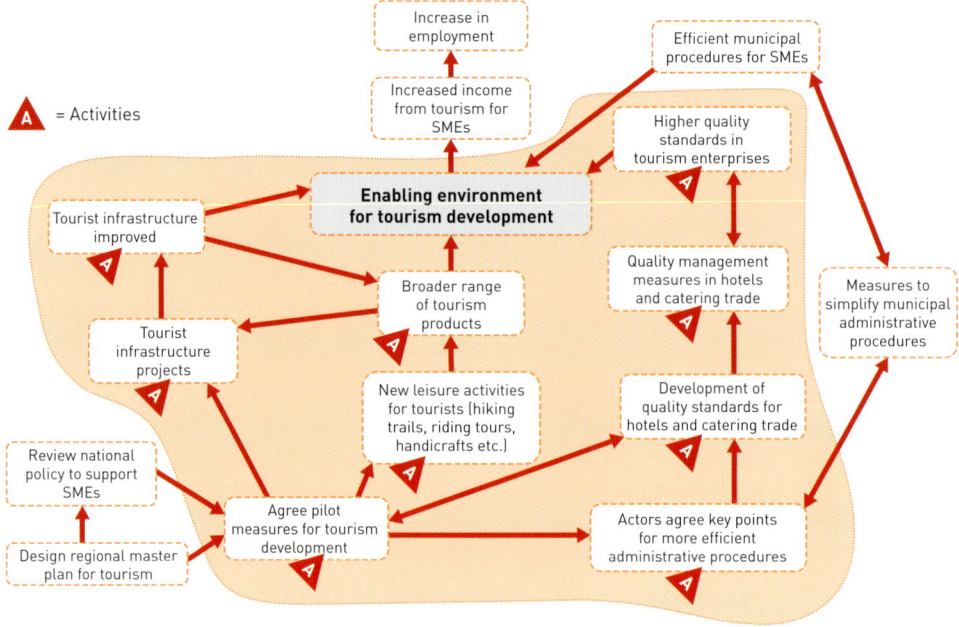

Figure 6: Example of a results model for tourism development

- Improve infrastructure (repair and maintain rural roads, repair bridges, secure viewing points, signs etc.).
- Broaden the range of tourism products (hiking trails, riding tours, handicrafts, typical food etc.).
- Raise quality standards across the board in hotels and the catering trade.

Here we see that the joint results hypotheses underlying these points is that conditions for tourism development will become more conducive if progress is made in these three areas. This would be manifested in higher turnover and higher employment figures. Two further areas – a review of the national SME policy and streamlining of administrative procedures in the municipalities – are being pursued by the ministry of economic affairs and local authorities outside of the project. The boundary between the areas addressed within the project and external initiatives is demarcated by a joint decision of the cooperation partners regarding the management of scarce resources. The lines of action by the project are then assigned activities, at which point the results model becomes the basis for operational planning.

Each of these lines of action contains results hypotheses that build on each other. For instance, it is assumed that the agreements concerning pilot measures for tourism development will lead to the planning and implementation of infrastructure projects. In some cases lines of action also intersect. Implementing quality management measures in hotels and the catering trade, for instance, will lead to higher quality standards in enterprises. Conversely, this will make enterprises more willing to agree additional measures for quality management. Finally, these results will combine to improve conditions for tourism development. This in turn will create an enabling environment for SMEs to increase their business in the tourism sector, thus causing employment in the region to grow. Neither of these changes, though, was facilitated directly by activities within the project.

This illustrates how the results model is derived from a joint objective. In this context it is helpful to make explicit the results hypotheses for the intended changes in the area of social concern, and to define the objective precisely. The results model brings to light the dynamics of the change process within the sector or social sub-system. This is to say that it describes (i. e. visualises) the inputs to be generated by the cooperation system, the needed changes among the various actors and the ways in which these interact. The model can then be discussed. Actions are agreed to help bring about the changes. Milestones can also be defined to indicate during implementation whether the cooperation system is moving down the right path.

The model helps the cooperation partners to talk about all these issues and negotiate their respective contributions. Communication of this kind then forms the joint starting point for management within the cooperation system, which involves continuously obtaining relevant information, and monitoring whether the intended changes are actually being achieved.

How GIZ understands objectives and results

GIZ understands results as the intended or unintended, positive or negative changes in a situation or behaviour as the direct or indirect consequence of an intervention.

The module objective (outcome) for technical cooperation (TC) measures represents the result that can be achieved – realistically and on the basis of a binding commitment – within the time frame and financial scope of a proposed intervention. It describes an intended, defined effect that an intervention will have on the target group, (regional) public goods, structures or policies.

The results model shows a progressive sequence of causally interdependent positive results. It depicts a change process that GIZ influences through a project. Since it has been agreed with the cooperation partners, the results model is compatible with both the project and the partner structures and processes. It is helpful to ask the following questions when preparing the results model:

- Who or what needs to change so that we can achieve overarching goals?
- What are the strategic options? Where will we define the project objective together with partners and commissioning parties?
- Who will we cooperate with?
- Which responsibilities will GIZ have towards the commissioning party?
- What will GIZ's contribution be toward achieving results, and what instruments will we be using?

For commissions placed by the German Federal Ministry for Economic Cooperation and Development (BMZ), the key points agreed are then transferred into a results matrix. ●

Results-based monitoring – a navigation tool

The cooperation system could be seen as a black box – particularly in view of its fundamental unpredictability. The only knowns would be what consulting, training and financial inputs were being fed in. At the other end it would be possible to measure for instance whether certain services had become better. This sounds not only pragmatic, but also practicable. But would it also be helpful? Only to a limited extent, because we would have to wait a long time before the black box produced results. And what if the results were negative, after long years of work? Or what if the results were positive, but no one was in a position to know whether they could actually be attributed to the project inputs?

In that case it would be better to take a look inside the black box to see how it works – i.e. how results hypotheses are formulated. Although this is more complex it does allow cooperation systems to continuously review the validity of the objectives formulated at the start.

In cooperation systems the steering structure is responsible for continuously supplying decisions. At the same time it ensures periodic comparison between the baseline and the changes achieved, from which strategic conclusions can be drawn. Professional cooperation management requires decisions to be made not arbitrarily, but on the basis of sound information. For this purpose the cooperation system requires a kind of feedback mechanism.

Ultimately this feedback will involve systematic results-based monitoring. This should supply information on the results achieved so far, support the monitoring of possible risks and capture the perspectives of the actors involved concerning the project. A monitoring system performs precisely this function: it supplies the cooperation system with feedback on its own effectiveness, enabling it to take appropriate decisions.

One key requirement for a successful monitoring system is that the agreed objectives should be operationalised using indicators. This ensures that the objective is clearly described. Appropriate indicators are obtained by systematically seeking answers to the following question: 'What will be different if we succeed here?' Indicators are parameters. By regularly collecting information on these parameters it is possible to measure whether the cooperation system is developing in the right direction. As well as a verifiable baseline value, indicators also always specify a verifiable target value.

Using the results model it is also possible to identify the risks that may jeopardise the achievement of objectives by the cooperation system. The monitoring system should therefore also be used to monitor the risks. Here too, indicators help describe the risks in ways that all actors can understand, and measure them in practice.

Indicators

Returning to the example of tourism development, the initial baseline situation for the line of action 'New recreational offerings' might look like this: The region currently offers two hiking trails, one provider of riding tours, and a village offering walking tours and visits to the local carving workshop. One indicator measuring the result 'New recreational offerings' could then for example be: 'By the end of the year the number of documented hiking routes in the region has risen from two to four, and has been supplemented by a well-signposted bike and walk trail.' This target value would reflect what had been agreed between the actors. This means that here too a crucial role is played by the willingness of the cooperation partners and the resources available to them. Of course, it would also be desirable to have six hiking trails and four bike and walk trails, as well as more providers of riding tours – guided donkey tours might also be a good idea for creating new and appropriate leisure offerings. Yet if the right actors are not (yet) on board, an indicator of this kind could not be achieved and would therefore be unrealistic.

As well as providing information on the achievement of objectives, monitoring should also supply information for day-to-day management, and bring to light 'blind spots' explaining why and how results have (or have not) been achieved. What are the project's strengths and weaknesses? To what extent have specific actions already been implemented? Have certain milestones already been reached? In other words, monitoring is a key basis for well founded decision-making within the steering structure. If the monitoring results deviate very significantly from the target values, this is an important indication of the fact that the original hypotheses should be thoroughly scrutinised.

Results-based monitoring thus provides the actors involved with the basis on which to review their respective contributions to the cooperation system. Sometimes they will need to make adjustments within their own organisational context in order to be able to deliver their specific contributions. Sometimes monitoring will document success stories that vindicate the decision taken by the actors to get involved in the cooperation system. Ultimately, the only way to navigate the complexity of social systems on a reasonably objectives-oriented basis is for the cooperation system concerned to continually monitor its own results.

Capacity WORKS in results-based monitoring

The monitoring system supplies important information on the 'what', i.e. the achievement of objectives in cooperation systems. This is important for measuring success in the system concerned. However, it is often the case that projects in particular, i.e. temporary cooperation systems, are commissioned by third parties to bring about certain changes. In this case there is also the issue of accountability to those commissioning parties, to whom reports will be submitted providing information on the objectives-oriented use of resources.

Making social changes sustainable shifts the focus onto the 'how'. This is to say that as well as a number of meaningful and quantifiable indicators for measuring results, the cooperation system also requires a qualitative assessment of its own development. The five success factors of the management model Capacity WORKS provides a good framework for this very purpose. Each success factor covers a key perspective for successful cooperation management.

Capacity WORKS in monitoring

Returning once again to the example of tourism development, the actors involved might structure their discussion in relation to the questions below. These always have one eye on the permanent cooperation system and the other eye on the project:

1. Strategy
What strategies for economic development and employment promotion can we tap into?

Is our objective sufficiently clear and does it promise to deliver benefits to all the actors involved?

2. Cooperation
Have we got the right actors on board? Or are there actors outside our project who we should involve more closely (e.g. environmental agencies and environmental NGOs, tour operators or national SME development institutions)?

Are our forms of cooperation the right ones to ensure that both the local public administration and local entrepreneurs feel that the logics by which they operate are being taken seriously?

3. Steering structure
Have we taken sufficient account of the decision-making structures within the municipal administrations and the local chambers of tourism, and the programmatic decision-making of the provincial government?

Is our steering structure efficient enough to reflect the needs of the actors involved while at the same time guaranteeing sufficiently brisk decision-making?

Does the steering structure make use of the data supplied by the results-based monitoring (RBM) system in its decision-making?

4. Processes
Which processes within the local government administrations need to be adjusted in order to improve service delivery to SMEs? How do the provincial government's budgetary decision-making processes need to be adjusted?

Is the project feeding the results of results-based monitoring back into decision-making to update the project strategy?

5. Learning and innovation
How are tourism businesses and local governments currently responding to customer feedback regarding improvement of the services offered?

How will we identify the learning requirement with regard to standardisation, improved cooperation, capacity development for specific organisations and training needs?

The key questions contained in the five success factors (see the section 'The success factors – an overview') can be used to help identify the right problems. Under some circumstances it may also be helpful to formulate indicators here so that target values can be agreed which the actors involved see as leading to a successful outcome. These indicators will provide the cooperation system with information on the quality of cooperation.

Alternatively a common option is to agree on a yardstick for these 'soft' issues that provides broad guidance, along a scale such as 'everything okay – room for improvement – urgent action required'. It is then a straightforward matter to identify next steps or document success stories.

The more systematically the perspectives of the five success factors are integrated into cooperation management (which includes integrating them into the monitoring system), the better able the cooperation system will be to identify efficient solutions for the challenges faced.

Seen in this light, a monitoring system performs various functions. It generates information that is relevant for decision-making and reporting. Information on effectiveness is also useful in PR work. Monitoring brings to light areas of learning, and enables lessons learned to be harnessed for future learning processes. Resources can be steered more efficiently once it becomes clear where closer attention is required, and where less attention will suffice.

Effectiveness for sustainable development is one of the key yardsticks of quality in GIZ's work. Managing for results is the cornerstone of all GIZ projects. This means that any project needs a results-based monitoring (RBM) system to provide information on its effectiveness at any time, and supply proof of its results. Monitoring data are used for steering, accountability and learning.

GIZ's RBM system involves a two-pronged approach: measurement of results using indicators, and the KOMPASS procedure, which uses unstructured interviews. Indicator-based RBM is derived from the project's results model. It involves monitoring the change process depicted therein in relation to quantitative and qualitative indicators. This provides the team and the partners with regular pointers as to where the project stands in relation to the intended results and within the planned process. The KOMPASS procedure uses qualitative methods and tools to systematically survey the opinions and experiences of various project stakeholders (e. g. target group, cooperation partners) on a predefined question or problem. This open approach is designed to bring to light blind spots in the results model, and question them.

The RBM system encompasses the following elements; results hypotheses, assumptions and risks, objectives and results indicators and possibly specific indicators on important crosscutting themes (e. g. gender equality, poverty reduction, environmental protection and natural resource conservation, peace and security).

Before new quantitative or qualitative indicators are developed we have to determine whether the partner institutions already have monitoring systems that could capture the intended changes. A data collection plan is drawn up based on the indicators to be measured and monitored; data are continuously collected, evaluated and analysed. These data indicate the degree to which the intervention has been implemented and the results achieved, and bring to light any possible need to adjust strategies.

The RBM system thus helps the steering structure reach management and strategy decisions systematically. Once a project has been completed, monitoring data then become key elements of the evaluation process. They enable us to substantiate results, assure quality and meet our accountability obligations.   ●

Capacity WORKS in cooperation system management

The five success factors supply the perspectives that are important for describing the status quo in an area of social concern: What strategies exist? Which actors are important? How are decisions taken? What are the key processes? Where are capacities needed, and which ones?

Once this initial situation has been analysed, objectives can then be identified. These describe the 'what'/'what for'/'reason why' of a cooperation system. The results hypotheses that postulate how these objectives can be achieved are made explicit in the form of a results model. This enables all the actors involved to develop a shared picture of the change process.

When designing the strategy for change, the five success factors help make tangible the 'how' of achieving the objectives in the cooperation system: Who will cooperate, and how? How will decisions be taken? Which processes are important? How will learning be managed?

Appropriate indicators are formulated both for the results and for management within the cooperation system. These are then incorporated into a monitoring system. The system supports continuous monitoring of the results, the risks and management in the cooperation system. It then supplies information on the basis of which decisions can be taken in the steering structure.

When implementing specific activities, the five success factors supply appropriate materials and tools to help orient the management of the cooperation system toward the objectives and results. To keep making sure that the path taken is still leading to the objective, the actors involved periodically subject the objectives and results hypotheses to close scrutiny.

The monitoring system supplies information on any adjustments that may be needed in order to continue working effectively. In the following chapters we will now describe each of the success factors in detail. We will explain the highly specific contribution made by each success factor to the management of cooperation systems for sustainable development. It will then become clear to the reader how closely each success factor is linked to the others.

Depending on their requirements, the actors (such as those in the tourism development example) will in due course focus selectively on one or other of the success factors. The information generated by monitoring the results will suggest which success factor the actors should take up for closer consideration. In our example questions like the following might come up: Is the change process deadlocked in the local administrations because the councillors are refusing to collaborate? Should the councillors concerned be more closely integrated into the steering structure? Does the system need an additional process or activities in order to persuade them? Do top administrators and mayors require support from the chambers of tourism, and does the system need to establish a corresponding cooperation process?

Reading the following chapters will help you to ask the right questions. In this context 'right' means 'useful', i.e. useful to real actors in real cooperation systems with their very own objectives.

Success Factor – Strategy

Motto: Negotiate and agree on the strategic orientation

Actors often accept compromises when they take strategic decisions in a cooperation system. These decisions usually do not fully meet the interests of the various actors. Yet they often have a more long-term and fundamental effect than the actors responsible would have assumed when they actually took the decision.

Let us cast our minds back to 1947. Most European countries were in a state of economic and political ruin. Their populations had been traumatised by a devastating war. In this situation the American Secretary of State, George Marshall, made an extraordinary proposal regarding the system of cooperation between the allied states. This was designed to have a lasting effect on cooperative relations within Europe. His idea was that the USA would provide an extensive investment package to boost the dormant European economies and help them get back on their own two feet.

The plan was extraordinary because it broke with several familiar patterns. The attitude with which the victors were treating the vanquished after the war was a new one. Marshall's proposal differed sharply from other plans for post-war Europe that were dominating the debate in the USA at the time. These included the ideas put forward by the US Secretary of the Treasury, Henry Morgenthau, who advocated transforming Germany into an agrarian state.

Also new were the premises and values underlying his answer to the question of how power relations in the second half of the 20th century could be stabilised in order to foster peace. He believed it would only be possible to reconstruct Europe from the ruins and create a continental system of cooperation by deconstructing the power of national interest and independence.[2] Those who supported Marshall's proposal had recognised that the individual nations were no longer able on their own to deal successfully with the challenges created by the war. The dramatic situation in the immediate aftermath of the war thus smoothed the path for a slow but sustainable restructuring of the European continent and transatlantic relations.

The national discourses which this prompted were controversial. The USSR did not support the Marshall Plan. After war had raged across its territory for four years, the economy of the USSR lay in tatters. Sixty million war dead placed a heavy burden on the country's dealings with the defeated Germany. Moreover, the Soviet post-war project did not provide for the establishment of a market economy. The Communist movement was pursuing different interests and values. During the war the cooperation system of the Allies had been based on the joint objective of defeating Nazi Germany and its allies. Once the common enemy was beaten, this objective proved to be no longer capable of supporting cooperation in the post-war period. This cooperation quite literally fell to pieces.

There was also resistance to the proposal of the American Secretary of State in the US Congress. Many members believed that the planned investment in faraway Europe would weaken the USA. Given the fact that the USSR was extending its power base, this seemed all the more threatening. Yet this very threat of communist influence, plus the military interests of the two major powers – the USSR and the USA – ultimately persuaded those concerned that the Western European states and the USA were dependent on each other.

Marshall's idea for the new cooperation system in the West ultimately won the day. And so the Marshall Plan, as it was known, was born (the official name of the American investment programme was the Economic Cooperation Act of 1948). By 1952 the USA had made a total of USD 12.4 billion (equivalent to around EUR 100 billion today) available to it. Within this cooperation system the Organisation for European Economic Cooperation (OEEC) was established. This organisation put forward a four-year recovery programme, and assumed the role of a European supervisory body for the impending reconstruction process. These decisions had positive consequences for what was to become the Federal Republic of Germany, whose representatives were not directly involved in the negotiations. The reconstruction process coordinated by Europeans and funded by the USA meant a continuous reduction in the size of reparation payments. This defused the revanchist tendencies that had emerged between the two World Wars.[3]

When we consider this version of history, we ask ourselves: Was it really like that at the time, or are there are also other perspectives on the events of the day? In many of us this example will cause an emotional response, prompting us to agree or disagree. Yet when we look at it as an example of political cooperation, the question is not how things 'really' and 'truly' were. Because regardless of whether they were US American citizens, French citizens or citizens of another country – everyone was affected in some way by the history of this cooperation system. And each and every individual may give a different description of the motives behind the strategic decisions taken at the time.

Yet this leads to a key insight, namely that developing the strategy for a cooperation system is a difficult task. Ultimately the strategy must ensure at least for some time that the actors agree on important points and are willing to take joint action. Sometimes they succeed in agreeing on basic values and assumptions concerning their shared context. Usually, though, the interests of the actors involved remain as distinct as they were before. Working on the strategy enables them to identify overlaps in their different interests, and take joint action.

Let us return to our example. In retrospect it may appear that there was 'no alternative' to the Marshall Plan. The fixation on national solutions was broken down, and the path to a joint understanding of capitalism, democracy, cooperation and economic aid was opened up. This conclusion is a false one, however, because the logic of cooperation systems is different. The responsible actors had to overcome any obstacles in their own countries, which included fighting hard political battles, some of them with each other.

A glance at the history of the peace treaties suggests that quite different strategies might have won the day in the wake of the Second World War. Though it may have been in the interests of some actors to contain the influence of the USSR, while others were hoping for a peaceful Europe, yet others were seeking to create an economic environment more conducive to capitalism. As important as these differences may be when we try to understand the actors involved, the outcome remains the same: all the actors played their part in strategically orienting the cooperation system in this direction, a fact that continues to have an effect today.

The specific perspective of the success factor 'strategy'

The success factor (SF) strategy focuses on how actors consciously negotiate and decide on the strategic orientation of their cooperation system. Like the other success factors in Capacity WORKS, it is highly effective in cooperation systems when the actors understand that the apparent nonchalance involved is deceptive. Strategy appears to be self-evident, otherwise (so some would ask) how could people work professionally?

However, on closer inspection we see that strategy touches the nerve centres of a cooperation system. Strategy asks the actors to share their deep assumptions and convictions. Strategy makes these assumptions explicit, thus inviting others to object or contradict. This generates fresh impetus as different interpretations of reality appear alongside each other. Strategy invites the participants to develop ideas on which joint actions can be based. Strategy provides an arena in which the actors involved can note their differences calmly, and ask themselves what future they might be able to share. Once the actors have the courage to admit to themselves that strategy is not self-evident, and enter into this arena, entirely new horizons of possibility open up.

Working with the SF strategy develops capacities within the cooperation system. Depending on the particular design of a strategy it can be used to attempt to influence societal frameworks. When developed and implemented jointly, strategy strengthens the cooperation relationships among the actors. The organisations taking part learn how to feed their own perspective into the joint work on the strategy. The individuals involved develop their competencies for implementing the processes needed to design and operationalise a strategy, and review it on the basis of the results it generates.

Paradoxes in strategy work

Everyone is talking about strategy, but what exactly is it? Strategy is a buzzword. It refers to a core economic competence, one that is important for organisations which operate on markets and wish to define their position there. Strategy work looks at how organisations satisfy the needs of customers, clients or stakeholder groups. In business, good strategies improve turnover, profits and reputation; in societies they improve the relevance and impact of organisations' responses to problems. These are the criteria by which the success of strategies is measured.

Organisations need strategies, regardless of whether these emerge over the course of time or whether they have been formulated explicitly. Through strategy work, organisations look at how they can best adjust to their environment and retain their position in the future. How they deal with the future varies widely from organisation to organisation. At one end of the scale the response is for organisations (or cooperation systems) to abandon themselves to fate. This leads to a rejection of any kind of strategy work. The other end of the scale is the arrogant belief that absolutely everything can be planned and made to fit. Between these two poles, cooperation systems and organisations must deal with the following paradoxes when working on their strategy:

- The future is unpredictable and uncertain – yet even so, we respond to it by planning.

- We respond to the variables of the future that cannot be influenced with the expectation that our actions will be effective.

- The unknown future always entails the need to change. This triggers anxiety and insecurity among the actors involved. To overcome with this they must respond with courage and optimism.

- Statements about the future are always to some extent based on knowledge of the past (it was Henry Mintzberg who pointed out how risky it is to keep your eyes on the rear-view mirror when driving a car).

- When actors experience a sense of paralysis, like the rabbit gazing at the snake of future developments, they must respond by reflecting and then acting, rather than losing their heads and avoiding the issue.

Strategy is all about dealing with these issues. For a cooperation system to be capable of acting it must discuss two sides of things: the more emotional side dealing with the imponderable issues, and the planning and conscious management side of things that appears to be objective. Strategy work creates a basis for continuous decision-making even in the face of uncertainty.

Strategy increases the ability of the cooperation system to act

More so than businesses, cooperation systems operate in the social sphere. They do not deal in the currency of profit, but aim to achieve results in areas of social concern. They aim to make a contribution toward efficiency, values, norms and meaning within a society. One thing that is crucially important to the actors within a cooperation system is the question of how they can remain capable of acting in the future.

Cooperation systems lack both the boundaries drawn by organisations, and the decision-making principle of hierarchy, both of which can make it easier to define a strategy. When actors join forces to pursue an objective in an area of social concern, they will have some very different ideas when they first begin cooperating. In most cases, all the relevant actors are not yet sitting around the table. The cooperation system must shake itself into shape, which it does by working on its strategic orientation and developing its capacity to act.

Strategy as an orientation and a process

When strategy succeeds in taking shape, it is manifested as a 'pattern in a stream of decisions'[4] – at least according to Henry Mintzberg's definition of it. Within a cooperation system, the strategic orientation of the system and that of the organisations participating in it must match each other. This kind of pattern in the stream of decisions can only arise if and when the various actors agree to negotiate one or several objectives with each other. This willingness has consequences, because the strategic orientation affects not only the cooperation system, but also the strategies of the organisations involved.

Within organisations, strategy development can be performed in various ways. It can be a planned exercise by the top management, it can evolve at various points within the organisation, it can be based on external expertise, or it can be developed jointly[5]. All these variations have advantages and drawbacks. In cooperation systems things are different. During the initial phase at least, the strategy should be developed in a joint process.

The process involves various steps, all of which are equally important: (1) analyse the current situation, (2) devise options, (3) decide on an option, (4) elaborate the strategy, (5) integrate the strategy into operations. If the actors omit one or several steps because they believe that sufficient clarity already exists, then they miss an important opportunity. What they miss is the opportunity to engage with each other. Although this may sometimes be difficult, it does allow the actors to deal with each other honestly and develop a joint perspective that is realistic. Moving quickly to achieve harmony and selecting the most obvious option is a risky way to proceed, because it can create a false impression that everyone is in the same boat and there are no alternative solutions. The SF strategy therefore shapes the spaces for communication that allow this discussion to take place.

A joint commitment to objectives

Capacity WORKS supports professional cooperation management from a variety of perspectives. The SF strategy focuses chiefly on the issue of which objectives and results the actors wish to achieve: What are the right things that we should be doing? This question always goes hand-in-hand with the next question: And what would be the right way to do those things?

The second question asks which path should be taken to achieve the objectives through specific actions. Conclusions drawn from the discussion of the objectives (the 'what'), of course also affect the path for change (the 'how'). Conversely, the quest for possible paths for change may also place objectives in a new light.

The cooperation system becomes more firmly established when it reviews its strategic orientation from time to time. The system as a whole, and each individual actor in it, then become part of a joint logic of action that allows the aforementioned 'pattern in a stream of decisions' to visibly take shape.

What strategy does

Both the process and its outcome, a properly formulated strategy that is transparent and clear to all stakeholders, do several things:

- They help the cooperation system to do the right things.
- They clarify the actors' expectations of the cooperation system and of each other.
- The resources and capacities that exist within the cooperation system are put to efficient use.
- Actors in the cooperation system are motivated to pursue its objectives.

- Participation in the process and the clarity of the results encourage the actors involved to commitment themselves to these objectives.

- The strategic approach is seen in terms of the future, and becomes detached from the limitations of the past.

- The actors know which paths they are pursuing in order to achieve the objective, and which paths have been abandoned.

- The joint activities target existing potential for social change, thus ensuring sustainability.

- The actors possess leeway to take appropriate action within the framework of the strategy.

The strategy development process

Detailed descriptions of the possible steps involved and the instruments that can be used when developing strategies are available in the literature written for businesses and organisations. Much of this can be applied to cooperation systems, including for instance the distinction between normative, strategic and operational management[6].

However, the approach to strategy development in cooperation systems must differ from the approach taken in businesses in at least two respects. First of all, cooperation systems are always part of the solution (and in that sense part of the problem) in social sub-systems. They are models of the results that they wish to achieve in the sector

Consider the following example. An advisory project aims to support a ministry of economic affairs in a European country in making better use of the European Union Structural Funds to promote competition and innovation. The project is focusing on transparent and efficient tendering procedures. The following actors will be involved in the cooperation system to be established for this purpose: the ministry itself, its agencies, the ministry of financial affairs, the enterprises concerned and universities.

Relations between the actors may be characterised by mistrust, mutual accusations of incompetence and corruption. The discussion of strategic issues can only succeed if the patterns of the behaviour that have so far prevailed are broken down. The cooperation system itself is part of the problem, the change process and the solution. New spaces of communication need to be created in which sensitive issues can be discussed and trust established. Here we need to identify points where success can be achieved on a small scale.

As in management within organisations, the inner core of the cooperation system will also be very closely observed by the other actors in the relevant social sub-system. These actors will base their judgement of the cooperation system on the actions of the core in general, and the results produced by the system in particular. If there are no success stories, or if these are not visible, then no model will emerge to inspire the trust needed to work together on more ambitious objectives. In this case strategy development must first focus on minor but visible successes, in order to then initiate broader change processes.

Secondly, strategy development for cooperation systems differs from strategy development for businesses in that the former revolve around social issues that often also have a political dimension. In this context, tools and strategies developed purely on the basis of the logic of the marketplace will not work. Cooperation systems must seek to analyse and understand their environment using approaches drawn not only from the economic sciences, but also from the political and social sciences.

Strategy is a creative throw of the dice, a bold and hopefully encouraging glance into the future. The actors involved should not allow themselves to be guided by the constraints of the present. A coherent strategy cannot be identified simply by following a clear sequence of questions and steps. The paradoxes mentioned above will block the path. The actors plough the common ground by sharing ideas on relevant issues and formulating hypotheses. As they move around the cycle the strategy becomes more coherent and tangible. The 'strategy loop'[7] shows how the steps fits together in the feedback process, while at the same time paying tribute to our way of thinking, which still remains relatively linear.

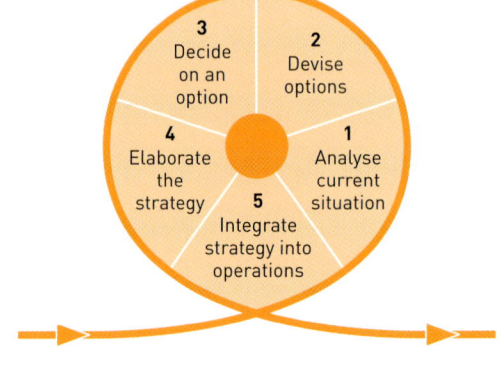

Figure 7: The strategy loop

Analyse the current situation (1)

It is not possible to develop a strategy when surrounded by the turbulence of day-to-day business. To enable the actors in a cooperation system to jointly adopt a new perspective, they need a fresh standpoint in the truest sense of the term. The actors must leave the 'stream of operational activity'[8] in order to gain strategic altitude.

First of all they must analyse the social context of the cooperation system: What trends and developments are becoming evident? Which actors are determining events? What challenges exist, and what forces are at work? What social patterns characterise the behaviour of the actors? What incentives lead them to act? Which actors possess energy for change? Which processes work? This glance at their environment is designed to enable the actors to move away from old interpretations and open their minds to a joint perspective. During this step the actors in the cooperation system construct a shared reality that provides a basis for all their further thinking.

The analysis turns the focus not only outward, but also inward: What potential, capacities and resources does a project have that will enable it to initiate changes in the sector? Which actors can contribute to that?

When performing this analysis the actors involved are deliberately encouraged to believe they do not know enough about how the social sub-system and the cooperation system work. This creates a feeling of never knowing enough, and therefore needing to persevere with the analysis. If the actors are not aware of this they could be tempted to turn the analysis into a kind of pseudo scientific research project. At the end, the complexity and the flow of data would be so great that the actors would then be paralysed. The world is full of strategies that have consumed a wealth of resources, but have ultimately ended up collecting dust on bookshelves. One of the challenges is to keep linking the analysis with the opportunities for action that are open to the actors. The information required is usually not lacking, as the actors possess a huge amount of implicit knowledge. Strategy work provides them with a stage on which this knowledge can be brought out.

The best guarantee of not getting lost in the analysis is to remember right from the start to condense the information. This means the actors should keep asking themselves: What does this mean for us and our objectives? If the answers to these questions are condensed into conclusions, by assigning them to categories such as strengths, weaknesses, opportunities or threats, this reduces the complexity appropriately. If the actors succeed in capturing the key results in memorable images then these will become part of the memory of the cooperation system, which all actors can plug into easily.

Devise options (2)

Limited resources create a compelling need for strategic thinking. Like organisations, cooperation systems also possess limited financial and human capacities. Consequently they must ask themselves which option will enable them to achieve the maximum effect. This means that options are pivotal in all strategic thinking. The results of the analysis will bring to light challenges, opportunities and strengths, as well as risks and weaknesses in the social sub-system. The actors will then develop various options for responding to these and gaining the appropriate leverage.

Here is an example: In Afghanistan, energy supply is to be improved in order to support rural development. Which paths might be followed in order to achieve this? In other words, what are the options?

Here is a selection of possible options:

- The national energy agency could establish small hydropower plants and pilot their operation.

- Research initiatives could be supported on appropriate sources of energy suitable for off-grid use in rural areas.

- A national policy for energy supply in rural areas could be developed.

- A fund could be developed to support provincial governments in implementing locally appropriate electrification strategies.

The example demonstrates that there are various ways of achieving an objective. These options may very well be unusual, or may appear unsuited to the context at first glance. However, creative effort is required in order to depart from the path that the actors have always trodden so well in the past. This is the only way to devise options that broaden their scope for action.

Options cannot be inferred from the analysis. They are rather the result of a creative act – they are 'devised', at the highest point in the loop, which is the furthest away from everyday operations. The actors allow themselves a brief period of time out. They are ambitious, and at certain moments almost think the impossible. They do not allow themselves to be constrained by potential limitations caused by a lack of resources or resistance, but nor do they drift into the realm of the completely unrealistic.

During this step, the possible traps are too little diversity, too little ambition and too modest objectives. To avoid these traps the actors need time to question the objectives in the cooperation system, or discuss the confidence placed in the cooperation.

Options contain fundamental and precise messages on what is to be achieved. They provide a clear focus. They contain information on the cooperation system, the partners who are to be involved and the contributions they are to make. They also indicate how the objectives are to be achieved. Coherent answers to these three basic questions make an option clear and convincing.

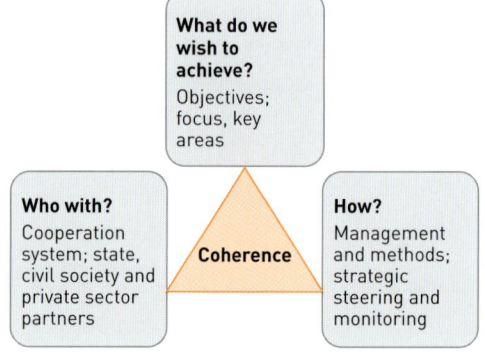

Figure 8: The triangle of coherence

Decide on an option (3)

The cooperation system must select one of the various options available. This requires careful discussion of the criteria by which this selection will be made. The actors must then evaluate the options. The process must be designed in a way that guarantees this. During this step, the trap is to favour one option too quickly and avoid the struggle to find the right criteria and perform a detailed evaluation. This step is complex, detailed and often arduous. Nevertheless it is worth insisting on, because each actor has to decide what they think is best. Only then will the decision be one taken by all the actors, and one to which they are all committed.

Elaborate the strategy (4)

Once the options have been evaluated the selected option is further elaborated. Sometimes this can also involve combining various options. The option is outlined as a vision of the future. The actors involved then recognise the scope of the changes, the key measures, the resources required and their own role. The latter is particularly crucial because the actors must communicate within their home organisations what obligations they have entered into. Developing a joint strategy in a cooperation system often also requires changes in the organisations involved, because this can call into question their strategic orientation, at least to some extent. One effect this can have is for representatives of organisations to stop short of entering into any binding decisions. In this case the process must be designed so that any open questions can be clarified within the home organisation.

Integrate strategy into operations (5)

This step links the strategic vision of the future with the practical operations of the cooperation system. The vision of the future now needs to be further differentiated using the other four success factors of the management model Capacity WORKS. What are the implications of the vision of the future for the forms of cooperation within the project? What demands will it place on the steering structure? What processes need to be established in the area of social concern in order to initiate the necessary changes? On which levels of capacity development does learning need to take place? These questions will guide the actions of the cooperation system when measures are being planned and resources allocated. This is where strategy turns into operational planning.

Conclusions

Strategy development is a demanding task, because it requires the actors to exchange ideas on their views of reality and develop a shared perspective. It may also be necessary to modify the agendas of the home organisations. The actors are required to consider options which they perhaps initially find disagreeable. They must agree on joint criteria for evaluating the options, and ultimately reach a decision that supports and is supported by the cooperation system. In other words, the decision and the cooperation system support each other. This is how patterns in a stream of decisions take shape.

Success Factor – Cooperation

Motto: Connect people and organisations to facilitate change

Modern societies face major challenges. Profound changes, often affecting several policy fields, have gathered pace. How can societies increase the use of renewables in their energy mix while at the same time ensuring that their economies remain competitive? What contribution can the education system make to facilitating access to the financial system for poor sections of the population, thus promoting economic development and social equality? Individual actors – whether from the public sector, private sector or civil society – cannot master these challenges alone. Increasingly, many issues require interventions that are agreed and implemented across national borders.

Moreover, modern societies are continuing to differentiate even further. New actors are emerging and articulating their particular interests. Competition for scarce resources is not a new phenomenon. Expectations concerning how the resulting conflicts can be resolved have changed, however. Whereas several years ago the state was still expected to reconcile these conflicting interests, today this is taking place increasingly through processes of negotiation between the actors concerned. The political participation of civil society actors plays a major role in this context. At the same time, the increasing specialisation within society means that more and more actors have to cooperate with each other.

Against this background more and more people are becoming aware of how important cooperation is. Since the term 'cooperation' has thoroughly positive connotations, it is tempting to draw the following (false) conclusion: 'The more people cooperate, the better. And the more actors that can be gained as cooperation partners, the more effective the cooperation will be.' This fails to take one key aspect into account, however. The individual actors must summon up energy for cooperation, which is a resource that we can compare with a scarce and expensive fuel. How much energy is to be consumed will depend on the task to be jointly performed. The more extensive the task, the greater the need for cooperation. Actors always have to choose the appropriate form of cooperation. A cooperation system is an option, and entails relatively close ties between the actors involved. Networks are considerably more flexible and entail less work for those involved. Consequently, anyone wishing to cooperate needs to carefully consider the implications. Actors will take a close look at the cost-benefit ratio of the cooperation.

When actors enter into their cooperation systems as partners, they nevertheless retain their identities. They still have to perform the tasks specific to their own organisation, and divide up their energies accordingly. To be able to understand and positively influence the dynamics of cooperation systems, organisations must ensure that they strike this balance between their own specific tasks and the joint tasks within the system. In this context we need to understand what an actor actually is. An actor is usually an organisation, or in some cases an influential public figure. Their behaviour is shaped as much by their interests as it is by their designated role and position within society. Actors are stakeholders; they are 'participants' in the process of social development that they try to influence. They are fundamentally autonomous in their decision-making and their behaviour.

Forms of cooperation geared to results help strike a balance between demands within the organisation and those arising in the context of cooperation. If this balance is not successfully struck, the scales will tip in the direction of the organisation's own interests. The representatives of all the actors involved do represent the interests of their home organisations, and act according to the specific logic of their respective organisations. They are accountable for ensuring that the cooperation is worthwhile from the perspective of their own organisation, and if it is not they will come under pressure. Professional cooperation management always attempts to minimise these tensions.

Conditions for the emergence and strengthening of cooperation relationships

When actors decide to enter into relationships of cooperation this does not in any way change the fact that they are autonomous. Since cooperation systems are always steered through processes of negotiation, however, actors must be willing to compromise. They will be willing to restrict the exercise of their autonomy if they expect to derive a major benefit from cooperating. For example, a government can only reform its education system if the other stakeholders involved (such as political parties, trade unions, and education and research institutions) can be involved in the process. In turn, the latter can only put their ideas on change into practice if they cooperate with each other and with the government.

Practical experience has shown that it is beneficial to adopt the **perspective of the individual cooperation partners**, and review the following conditions for the emergence strengthening of cooperation relationships:

- **Benefits:** The cooperating partners expect a benefit for themselves, and assume that they can only achieve it by cooperating.

- **Transaction costs:** The costs of cooperation are recovered through the results achieved.

- **The synergy rule:** The cooperation partners base their joint actions on the complementarity of their respective individual strengths. This is why they usually only accept cooperation partners who are able to create new potential through their strengths.

- **The fairness and balance rule:** The actors involved compare their own transaction costs and their benefits with those of the other cooperation partners, and react sensitively to any imbalances.

As well as the perspectives of the individual cooperation partners, the perspective on the cooperation system as a whole is also important. Are the actors willing to assume joint responsibility for a change process? Is the cooperation based on mutual appreciation? If the answer to these questions is yes, then something new can come out of the cooperation system that amounts to more than the sum of the contributions made by the individual actors. Professional cooperation management takes the following points into account right from the start:

- **Transparency of participation in the cooperation system and the roles of the organisations involved:** Workable cooperation systems require clarity as to who is part of the system and what role they play, who is not part of the cooperation system, and through what mechanisms actors can enter or exit from the system. This defines the system boundaries. Those who are part of the system are entitled to place other expectations on it than an external actor would be. At the same time, the cooperation partners expect each other to provide their respective contributions. Those who are not part of the system need not feel this pressure of expectation. If this transparency is not established, the system boundaries remain diffuse and the cooperation system loses its results-orientation. Practical experience shows that it is better to invest the necessary energy in clarifying affiliation and roles before beginning cooperation. This investment always pays out when implementing joint actions.

- **Orientation toward strengths:** The cooperation partners orient their approach toward their joint strengths that enable them to achieve the objectives they have set using their own resources. This first of all reduces dependency on external actors while at the same time strengthening joint identity within the cooperation system. Secondly it helps define realistic objectives.

- **Balance between cooperation and conflict:** Cooperation systems always contain some potential for conflict resulting from the individual and organisational interests of its members. Cooperation systems that work use conflicts between the actors involved by raising and addressing the different interests. If a cooperation system is unable to do this, the key challenge will be to strengthen the actors' capacities for dealing with conflict. Otherwise the joint objectives, and the actions designed to achieve them, will continuously be called into question.

- **Balance between strong influence (power) and weak influence (powerlessness):** Cooperation systems are never fully balanced. The circumstances of the organisations involved differ too widely for that. Professional and managerial expertise, financial resources, relationships and interests, and loyalties to other actors are just some examples. These aspects determine the extent to which the individual actors are able to influence the cooperation. Professional cooperation management makes these differences visible and uses them by addressing problems and identifying opportunities resulting from the differences in perspective. If the differences remain in the dark (even though all cooperation partners know of their existence, they interpret them completely differently), this will usually lead to a weakening of engagement among the cooperation partners. It is helpful to create special forms of cooperation and spaces in order to deal appropriately with issues of power. In some situations the actors concerned must be given an opportunity to address conflicts in protected spaces, without running the risk of losing face in public.

The specific perspective of the success factor 'cooperation'

The Capacity WORKS model supports the successful management of cooperation relationships. Basic concepts such as sustainable development, objectives and results, and all the success factors are based on cooperation. So why do we need a separate success factor (SF) for cooperation? What specific contribution does this success factor make which adds something to the perspectives of the other success factors?

Close cooperation creates a new social system. This system is defined by the joint objective, the actors involved, their relationships and the rules they set themselves. A cooperation system possesses boundaries that make clear which actors belong to it and which do not. In organisations, the influence of the founders can often be felt for a long time. Through the interaction between the actors involved, cooperation systems also develop an identity of their own that is shaped by their founder members. The SF cooperation focuses on how the cooperation relationships within a project can be managed as effectively as possible. This involves taking into account the relevant features of the permanent cooperation system, so that the project can generate its results there.

The SF cooperation analyses among other things the actors involved or yet to be involved. It reflects on their interests and attitudes towards the change goals of the project, as well as their influence and responsibilities within the sector. It also takes a close look at cooperative and conflictual relationships, and explores the scope for political participation. The decision as to which actors should be involved is always determined by the perspective of the objectives to be achieved in the cooperation system.

The aim is to determine which actors are relevant and must be involved, either because they are able to make an important contribution toward achieving the objectives, or because they can veto the change process. At the same time, objectives and results also incorporate the outcome of the analysis of actors (see the tool 'Map of actors'). For instance, if it emerges that the relevant actors do not support the envisaged objected to a sufficient degree, it follows from this that the anticipated results should be formulated more realistically.

Moreover, the SF cooperation focuses attention on defining the roles of the actors involved and using appropriate forms of cooperation. Both roles and forms of cooperation should be defined on the basis of an appropriate understanding of the area of social concern to which a project relates. Here it is important to recognise which defined roles and forms of cooperation a project can build on that have already been established, and whether these should and can be modified.

The SF cooperation also helps to continuously review the system boundaries of a project. Would other actors need to be integrated? Are the forms of cooperation appropriate? Are politically weaker actors also able to contribute their perspective? Have the roles been clearly defined, and are they respected by the cooperation partners? How are the cooperative and conflictual relationships between the actors developing? When addressing these questions it is important to remember that as the project unfolds, the system boundaries will be flexible and permeable. Changes in the wider setting often mean that a project must adapt, for instance by integrating new cooperation partners.

For cooperation to succeed, the necessary capacities must be developed on all levels. At the level of society, laws, mandates, cultural factors etc. create conditions that can enable or constrain cooperation between the various actors. While some organisations perhaps see it as their role to cooperate with external actors, others are more inwardly oriented and find it difficult to cooperate. The individuals who act within organisations are influenced by directives indicating the degree of openness that they may show toward other actors. An important role is also played by specific social competencies and practical knowledge on cooperation management, for instance concerning how to moderate meetings. The SF cooperation provides strategies and tools to develop the specific capacities of actors in this area. These capacities will then also be available for other contexts of cooperation outside of the project.

Internal and external cooperation

The actors who assume shared long-term responsibility for the joint project are located inside the system boundaries. This inner core of the project constitutes the cooperation system in the strict sense. This is the context in which the SF cooperation speaks of 'internal cooperation'. This is where all the relevant strategic decisions concerning implementation of the project are taken jointly. Moreover, the internal cooperation partners often also implement the planned activities. In most cases the steering structure of the project results from the division of tasks and roles for internal cooperation.

Beyond the inner core of the project there are often other, external actors with which the project cooperates. They do not assume responsibility for the success of the project. Nevertheless they are willing to support the project from outside, either on specific occasions or continuously, for instance by providing advisory inputs, financial contributions, political lobbying or PR work.

Relations of reciprocal exchange with these actors as external partners of the project can be built systematically, in order to mobilise their resources and thus generate related synergy effects. This will usually involve specific occasions that are of major significance for the project, for instance where a famous personality supports a public awareness campaign free of charge. The strategic challenge in cooperation system management is therefore to persuade the 'right' external actors to act as partners willing to exchange reciprocal benefits. These are identified on the basis of their profile, resources, knowledge, access to relevant stakeholders and other aspects. At the same time, the cooperation system must of course also possess qualities that make it attractive for these external partners, and be able to communicate them.

Over the course of time, the system boundary that draws a line between internal and external actors becomes blurred. Internal cooperation partners can become external partners, i.e. switch from being on the inside to being on the outside, if their role in the project changes. Conversely, an external partner outside the project can gain strategic importance and become an internal cooperation partner.

Cooperation systems and networks

So far we have been speaking for the most part about cooperation systems, whose objectives need to be negotiated just as much as do for instance the roles and contributions of the cooperation partners. The SF cooperation draws a clear distinction between cooperation systems and networks. Networks are not cooperation systems, because they perform highly specific functions and therefore also obey different rules. The distinction between cooperation systems and networks has far-reaching consequences for successful cooperation management. If these differences are overlooked, efforts to cooperate may fail as a result.

The actors involved select an appropriate form of cooperation depending on the objective to be achieved. If the actors involved for instance agree on a clearly defined objective, the cooperation then requires a high degree of mutual obligation and reliability. This is clearly illustrated by the example of the delivery of home care as a relevant social service. The cooperation partners are willing to obey joint rules and define the roles and contributions of all the actors involved. The cooperation system has a clear system boundary. Agreements are reached as to which individuals can receive home care services under what circumstances, and how these will be funded. Quality standards also need to be defined, as do mechanisms of quality control and sanctions in case of non-compliance with the agreed standards. In other words the cooperation relationships within the system are highly formalised, because the joint objective requires this.

One of the actors involved in the cooperation system might for instance be a welfare association representing the interests of the patients and their families. The association is very interested in lessons learned in neighbouring countries, because these may provide fresh ideas for its own work. Since this interest is also shared by similar institutions in the other countries concerned, a sharing of lessons learned then results, based on both face-to-face meetings and virtual communication. A learning and exchange network emerges, the purpose of which is to support the sharing of lessons learned among the parties concerned.

In this case, how strong is the need for binding rules? The aim of this network is rather general, and therefore the selection of members need not be exclusive. It is sufficient to be interested in participating in the network. If that interest wanes, or if the necessary resources including the time required are not available, the actors concerned can reduce their participation once again without being sanctioned. Networks of this kind derive their vitality from their openness, a small support structure, and a low degree of obligation that can extend to communities with no obligation at all. The formalisation of relationships – whether it involved rules of membership, the definition of contributions etc. – would increase the transaction costs for the parties involved to an inappropriate degree.

By not having binding rules the actors involved gain important advantages. The openness of network relationships is an inviting prospect because this kind of network can offer a very wide range of experience and perspectives. In this setting it is easier to discover creative solutions. The lack of obligation when cooperating means there is no pressure to take decisions, which would tie up some of the participants' resources that should really be used for open exchange. Moreover, learning and exchange networks of this kind are more flexible, can address new topics more easily and respond spontaneously, instead of having to go through formal decision-making procedures.

Furthermore, networks provide an opportunity to get to know other actors with whom formal cooperation relationships could be entered into if required. Networks transform potential relationships into actual relationships. A network also always offers a milieu for free-flowing creativity in which, unlike in organisations and cooperation systems, there is less pressure to succeed. Failures have barely any negative effects on the members not directly involved. Successes, on the other hand, are swiftly copied and always varied in new ways.

Networks are not per se precursors of formalised cooperation systems. However, if cooperation relationships between individual actors in a network become permanently formalised, this increases the likelihood that structures will be established – be it in the form of a cooperation project, an expert forum, a citizens' initiative or a business start-up.

In other words, cooperation systems and networks differ in various respects. The **purpose** of a cooperation system is to promote binding cooperation, whereas a network is there rather to support flexible associations. The regulated **membership** of a cooperation system contrasts with the spontaneous participation of actors in network activities.

	Cooperation systems	Networks
Purpose	Create obligation and reliability of contributions made by the cooperation partners	Exchange of relationship capital (know-whom: attractiveness and number of contacts), milieu with potential for future cooperation, sharing of experiences and ideas, joint learning and co-creation of practical knowledge
Membership	System boundary indicates who is a member and who is not; more flexible and permeable than in the context of a single organisation	Cannot be clearly defined, no fixed system boundary
Objectives	Agreed objectives that create the framework for binding contributions by the cooperation partners	Objectives tend to be vague, providing rough guidance for the contributions of the participating actors
Steering structure	Formalised steering structure, may be highly elaborate	Decision-making tends to be informal and ad hoc

Figure 9: Cooperation systems and networks – commonalities and differences[9]

The binding form of cooperation in a cooperation system requires an elaborate **steering structure.** By contrast, networks have informal mechanisms for taking decisions and managing their affairs. These differences are down to the different kinds of objectives which the two seek to achieve. The clarity of the joint objectives of a cooperation system requires considerably more structure than the more vague understanding of objectives found in a network.

If the structures required by the cooperation system 'home care' are underestimated, and supported using the logic of a network, the actors will probably not achieve their objective. Finally, the payment of contributions presupposes a minimum degree of obligation. By the same token, any attempt to formalise the network of welfare associations is in all probability likely to fail, because here the actors involved seek the benefits of creative exchange at low transaction costs.

Exchange between cooperation systems and networks

Projects are cooperation systems and therefore require structures. In many cases, networks can perform an important role for projects. Like external partners (who are involved because of the reciprocal benefits), networks can be involved for a highly specific reason and for a specific period of time. Cooperation systems can profit from the relationship of mutual benefit with a network. For instance, the cooperation system 'home care' might be interested in adopting the strategic approach of the international network of welfare associations. At the same time, the actors in the network might be interested in seeing their ideas influence the development of the cooperation system.

Both the actors in the cooperation system and those in the network then both expect to gain a possible benefit, without either of them having to change the way they work. The enabling principle of voluntary exchange, which is a characteristic feature of networks, is also evident here. The actors in the network are not under pressure to formalise their cooperation, nor do they need to assume responsibility for the objectives of the cooperation system. The cooperation system is not obliged to address in any particular way inputs offered by the actors in the network, and is certainly not obliged to take them on board.

If cooperation between a cooperation system and the actors in a network results in increasingly close ties, this may very well lead to a formalisation of relationships and an enlargement of the cooperation system. In this case, actors from the network would formally become part of the cooperation system.

Professional cooperation management promotes dialogue between actors and thus increases the scope for participation. If the stakeholders involved succeed in recognising their mutual dependency, formulating joint objectives and identifying appropriate forms of cooperation, change becomes possible.

Success Factor – Steering Structure

Motto: Negotiate the optimal structure

In cooperation systems the partners take all the necessary decisions together. What objectives do we wish to achieve? What strategy will we apply in order to achieve them? What specific interventions will this involve? These and many other questions must be answered if a cooperation system is to be capable of action. Unlike in organisations, there is no line manager who can resolve deadlock or take quick decisions. In a cooperation system the actors share this responsibility. The individuals involved must be able to distinguish between the logic of their home organisation, and the logic of cooperation, in order to be able to work effectively in both contexts.

Yet the principle of joint responsibility does not change the fact that cooperation systems are too complex to be able to guarantee that all actors are involved comprehensively and on an equal basis for all issues. In cooperation systems too, differences exist with regard to power and participation by specific actors. Sometimes it even appears that certain actors involved might be able to dictate or force decisions on the cooperation system. Yet when vital interests of other cooperation partners are affected, if not before, there is no way around the fact that decisions must be negotiated. Cooperation is based on the fact that the partners involved realise that they are ultimately dependent on each other, and are therefore willing to give up some of their autonomy in order to achieve joint objectives. This decision can in principle be revoked at any time. This happens particularly in cases where individual cooperation partners attempt to assert leadership claims, thus calling into question the overall basis of cooperation.

Once it is clear to everyone involved how decisions are taken, and what role the cooperation partners play in that, this provides a sound basis for continued cooperation. If some actors have the impression that the balance has been upset, this usually leads to conflict.

Many decisions on various levels need to be prepared and taken on a coordinated basis. The steering structure provides 'social spaces' for these processes of negotiation. Within these 'spaces' the partners agree on rules and roles, and continually take the decisions needed.

The following example shows how the different levels complement each other. The private sector and state actors within a planning region agree on a development strategy and a plan of implementation. These decisions are taken within a framework set by the provincial government through its development and financial planning procedures. This framework itself emerged from the dialogue between public authorities and interest groups, and refers to the policy directives formulated jointly by various ministries at the national level. A single central steering body would be unable to assess which topics required decision-making, and what the decisions should be. This means that the steering structure must be as elaborate or complex as the tasks it has to perform.

The actors involved are familiar with all these points because numerous issues are now dealt with through cooperation arrangements. The state would be unable to function if it did not cooperate with actors from civil society and the private sector. Businesses are often only able to meet the demands of the market if they seek a division of labour with other businesses. More and more often, civil society organisations are articulating the needs of their members vis-à-vis other social actors. In many countries they are assuming responsibility for tasks that have traditionally been performed by the state. Within a policy field there are as many steering structures as there are cooperation systems. Although these structures supply the decisions needed, they also tie up resources.

So as we see, there is no blueprint for ideal steering structures that would be universally suitable for cooperation systems in general. Each cooperation system needs to negotiate its own optimal structure. The steering structure enables the cooperation system to decide with appropriate speed, and transparently, which specific activities should be launched.

A project is a temporary cooperation system whose objectives and strategy relate to a policy field or parts thereof. This means the actors involved must take account of the steering structures that already exist in the policy field, and where appropriate use them. Otherwise severe friction may be created between individual cooperation partners, because parallel structures may violate important rules and increase the need for coordination unnecessarily.

Returning to the above example, it would for instance make no sense to create a round table for regional planning if a development council already existed in which the key actors were already involved. Furthermore, one purpose of a project is to act as a model for the permanent cooperation system. This also applies to the way in which the various actors within the steering structure negotiate decisions so that joint objectives can be defined and achieved.

Within the boundaries of the project it is then possible for example to try out new forms of participation that can subsequently be scaled up in the policy field. This prototyping can only succeed, however, if the actors involved assume ownership of the process. How far the impacts of this prototyping process reach depends on the willingness of the actors concerned. Capacity WORKS focuses on this central challenge through its success factor (SF) steering structure.

High demands placed on decision-making

The more complex the objectives of a cooperation system are, the more heterogeneous the group of actors usually tends to be. Their perceptions and behaviours are shaped by cultural, organisational and personal factors. A project is influenced by its environment, which is constantly changing. A technocratic interpretation of steering would be of little help here.

Complex cooperation systems cannot be planned on the drawing board. Nobody has them under control, and no one 'has a grip' on them. Although a good monitoring system is helpful, it does not guarantee control. Steering decisions have to be taken all the time, even though the actors involved know that these decisions may soon prove to be wrong, or at least in need of correction. Steering is therefore always an iterative process of observation, hypothesis formation, decision-making, implementation and self-critical review of the effectiveness of the decisions taken. The cooperation system is challenged not to lose sight of its objective, while at the same time keeping an open mind for fresh options. At the same time, though, the objectives set should be regularly reviewed and adjusted.

The steering structure is supplied with the information it requires for decision-making by the cooperation system's monitoring system. The monitoring system supplies the actors with relevant information for instance on the implementation status of the planned activities, the results achieved and relevant changes in the wider setting. In other words, the quality of the steering structure will depend among other things on how well the monitoring system works.

What the steering structure does

The steering structure encompasses all the planned and unplanned structures that arise in the cooperation system and are used by the actors to take decisions. In particular it sets out the rules, roles and responsibilities in the decision-making processes, and does the following things for the cooperation system. When strategic options are fed into the steering structure, the actors in it weigh up the options before taking strategic and operational decisions. Conflicts need to be recognised and dealt with in good time. Similarly, traditional management tasks such as resource management, operational planning, implementation management and monitoring also have to be performed by the steering structure.

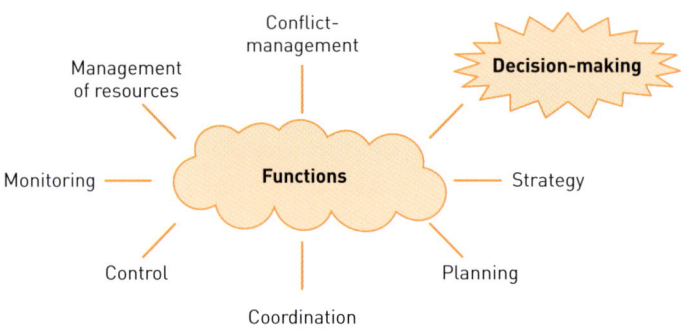

Figure 10: Functions of a steering structure

A cooperation system's steering structure is watched by all the actors involved, as well as actors in the wider setting. Many actors only ever see one segment of the cooperation system. By watching the steering structure they form a picture of how the cooperation system as a whole is doing. It would be naive to think that the actors within the steering structure only ever use the 'stage' that the 'audience' can see. Of course there are also informal opportunities and spaces 'backstage' where key decisions are often facilitated and taken. For example, it might be that municipalities in the aforementioned planning region find that decisions which would be important for local infrastructure projects are not being taken within the steering structure. This can lead to the mayors seeking alternative channels in order to influence the ministries. If these bypass solutions become frequent, this can erode the entire steering structure. This is why decisions and the pertinent decision-making processes must be communicated with the utmost care.

Unfortunately, we also often see how steering structures degenerate into a viewing platform for the mutual domestication of the cooperation partners involved. People meet, and we can then see how the individual actors carefully circle around each other, position themselves and gather information. After that they surreptitiously reach informal agreements with other cooperation partners, or return to their home organisations and gird up to pursue their objectives. This usually indicates that the fundamentals of joint cooperation have not been sufficiently hammered out, that the actors expect little from the cooperation, and that competition between them is high while trust is low. These tendencies need to be recognised early on and dealt with within the steering structure. If this does not occur the cooperation will presumably fail.

The way the steering structure is designed also has a crucial effect on the **quality of communication** in the cooperation system. This is where (among other things) access to information and the channels through which the various actors exchange views are determined. The agreed arrangements have a direct impact on the quality and content of the decisions taken. At the same time this strengthens the organisations involved and cooperation between them.

What is most crucial is that the actors who possess sufficient political power to support the envisaged social changes also participate. This is the only way to ensure that the project can serve as a model for scaling up in the permanent cooperation system. The more far-reaching a change objective is, the higher up in their respective hierarchy the representatives of the relevant organisation should be. If for instance a project aims to strengthen mechanisms of citizen participation for regional development, it must ensure that both civil society and state actors are involved in the steering structure. These actors can powerfully support the objective, and help ensure that the project and the change processes within the policy field match each other.

Levels of steering

Experience has shown that we should distinguish between politico-normative, strategic and operational levels of steering. This recognition of the subsidiarity principle for instance relieves high-ranking decision-makers of having to take decisions that can be taken by people at the next level down who have better access to the relevant information. This also ensures greater overall acceptance of the steering structure among the actors involved.

The example of regional development demonstrates the need for an elaborate steering structure: Actors within the region would not assume responsibility for the change process if all the decisions were to be taken by the top decision-makers in a few ministries. In many cases these decisions would presumably fail to meet the particular needs of specific regions.

The objectives and fundamental values and rules of behaviour during cooperation are negotiated and settled on the **politico-normative level**. The achievement of objectives is monitored, and any needed adjustments to the objectives are agreed. Fundamental conflict of interest or infringements of joint values are dealt with at this level. Returning to our example, the ministries of financial, economic and interior affairs reach agreement with the national association of local authorities to pilot an approach in two regions of the country. These organisations are represented by state secretaries and the president of the national association of local authorities.

The **strategic level** is a level at which the steering structure determines which path will be taken in order to achieve the objectives. The steering structure establishes an overview of and analyses progress made and deviations from targets in implementation work, reviews strategic options and agrees on milestones for further implementation. The directors general of the ministries, the executive secretary of the local authority association and representatives of the regional development councils perform these tasks and prepare a rough plan of action.

The **operational level** assumes responsibility for all the day-to-day decisions needed to implement specific measures within the prescribed strategic framework. This level provides the strategic level with a basis for decision-making by supplying information on progress and discrepancies in implementation. Representatives of the regional development councils as well as business associations, non-governmental organisations and municipalities develop detailed plans of action, and are responsible for their implementation.

In this context it is crucially important that the interfaces between the steering levels are well supported. The operational level, for instance, possesses an enormous amount of detailed knowledge that is important to the strategic level when taking decisions. At the same time, this detailed knowledge must also be made accessible in an appropriate form.

A clear understanding of roles and mandates, and mutual acceptance, are key to establishing close links between the levels of steering. The different levels of steering also use the information supplied by the monitoring system. The more complex the tasks of a cooperation system are, the more sophisticated and complex the steering structure will usually need to be. This means that several levels and 'social spaces' – in other words committees, working groups or workshop-type events – need to be distinguished and linked up with each other.

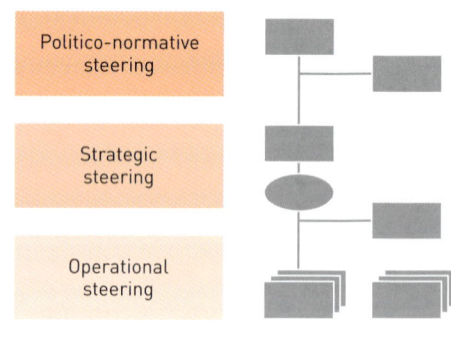

Figure 11: Levels of steering

The political and cultural context

Any project takes place in an established political and cultural context that is conducive to certain changes and opposed to others. This makes some things appear appropriate and expedient, while others appear inappropriate and nonsensical. In other words, people and organisations are not the only 'agents' that steer a project. Existing structures and conditions, ongoing communication processes and relationships, and established steering processes also steer the project. In other words, any project is embedded in a larger comprehensive system.

These existing organic structures in steering include, for instance, legislation, the administration or the private businesses in the market place. Steering and coordination mechanisms are defined in various ways and may be hybrid: important decisions can be reached through hierarchy, the market and/or negotiations between important actors. Ambient structures are often felt to be constricting and obstructive. Seeing things in this light misses the fact that these established structures have a very important role to play. They are repositories of past experience that provide actors with key points of reference when communicating and cooperating, and help them deal with each other in ways that are predictable and give them a feeling of certainty.

The specific perspective of the success factor 'steering structure'

The SF steering structure is dedicated to the issue of how a project is supplied with decisions. There are several reasons why the cooperation partners should be involved in steering the project. The steering structure is a pivotal link between the project and the permanent cooperation system in which the results should ultimately be achieved.

By providing detailed knowledge from various perspectives, the cooperation partners help ensure that decisions can be made on the basis of sound information. Participation in decision-making facilitates the development of new patterns of communication, as well as strengthening cooperation between the actors, also beyond the boundaries of the project. Transparent decision-making encourages the actors involved to assume ownership. From the above it is clear that designing and advising steering structures forms one of the core tasks of professional cooperation management.

The steering structure emerges through a process of negotiation which the actors involved can manage using Capacity WORKS. The process of negotiation does not come to an end once the steering structure is established, as if it were to remain fixed for all eternity. The structures continue developing through time, and should be reviewed in relation to the intended objectives and results, and if necessary modified. As time progresses the steering structure reflects developments in the cooperation project and in the context. It will pass through phases when it becomes looser (usually during critical phases of the cooperation system), after which it will be restructured and consolidated once again.

Many demands are placed on the steering structures of cooperation systems, yet ultimately there are only **two criteria for evaluating them**: 1. The steering structure must be functional with regard to the intended objectives and results. 2. It must be **appropriate** to the complexity and scope of the task.

We speak of **over-steering** when more resources would be required to establish and maintain the steering structures than to achieve the objectives and results. This is why it is so important to consider carefully whether and how existing steering structures can be used for a project. We speak of **under-steering** when there are many tasks to steer, but too few actors involved in steering them, who in any case communicate with each other only sporadically.

Every cooperation system develops a specific steering structure according to its particular requirements. It is therefore always a challenging task to jointly elaborate a functional and appropriate steering structure. In terms of capacity development, this requires more than just the relevant competencies of the individuals involved. Hierarchically structured organisations in particular must learn that the logic of steering in cooperation systems means making compromises, and that decision-making processes may sometimes be slow. Creating a model within the steering structure of a project that can be transferred to general cooperation between actors in the permanent cooperation system thus constitutes an important result.

The following examples show how steering structures reflect the specific requirements of the project:

Figure 12 shows an organisation chart of the levels and units of steering in a project. Its change objective is to operationalise regional development strategies. A supervisory board comprising top representatives of the ministries involved meets to formulate and review the project objectives. Experts from several universities in the regions form a scientific council to support the ministries in the form of expert discussions. The steering committee comprises top representatives of the ministerial directorates and development councils in the regions. This body also coordinates the operational level (sub-projects and support services). The coordination unit and secretariat, and the planning, monitoring and communication unit, are in permanent contact with the operational level. This level is formed by actors within the regions. This example shows a very clear structure that is typical of many cooperation systems.

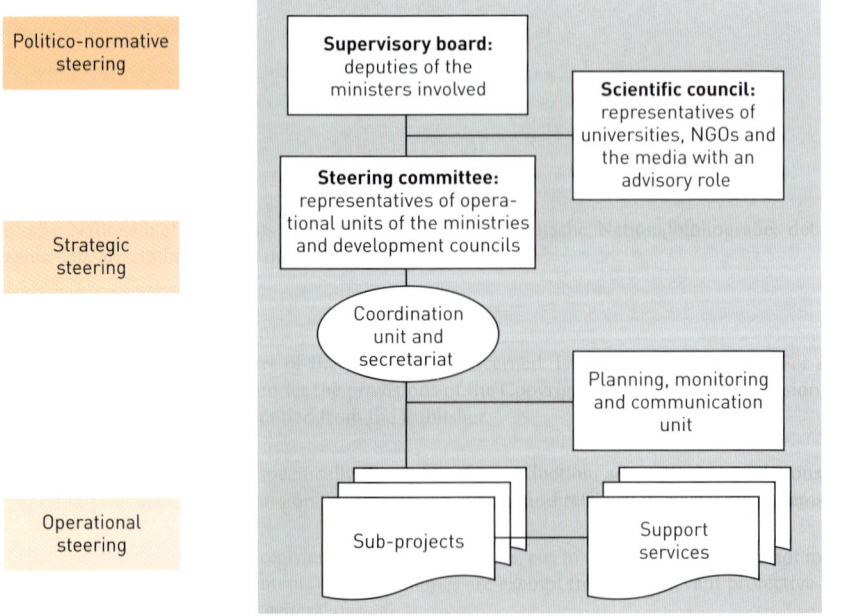

Figure 12: Formal steering structure with clear distinctions between the levels of steering

Figure 13 shows a steering structure functioning as a flexible project organisation. Unlike in the first case, no clear change objective has been defined. The project is mandated to develop proposals based on practical lessons learned that will subsequently serve as a basis for decision-making by the ministries. On the politico-normative level there is a steering committee responsible for taking decisions, comprising high-ranking ministerial representatives. This is flanked by two advisory bodies. A sounding board made up of academics supports the political debate from the scholarly perspective.

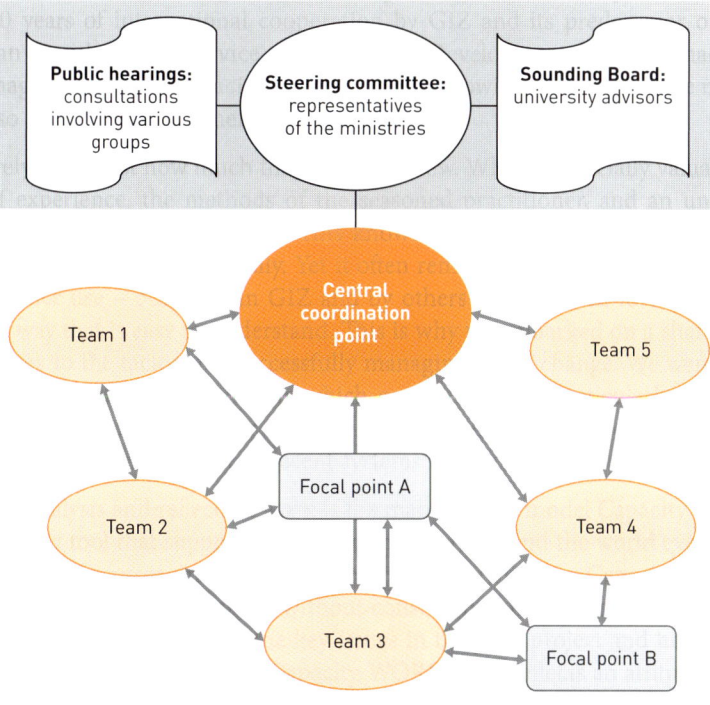

Figure 13: Steering structure with flexible project organisation

Public hearings are held at selected points in time to ensure that stakeholders from the regions are able to voice their perspectives. Both the sounding board and the public hearings need to be carefully designed in order to avoid any misunderstanding of the roles and to manage the expectations of the participants. Finally, the project is in a pilot phase, which means that the national government is not yet able to change public policies.

At the strategic level there is a central coordination point. This body mandates project teams and provides resources to implement and evaluate pilot projects involving for instance rural value chains, or new forms of citizen participation in development planning. Two focal points at the operational level are responsible for supporting the measures in the regions.

Compared to the first example this structure is considerably more flexible, though it also displays a greater need for communication. Partners whose organisations are more hierarchically structured must be willing to use horizontal forms of communication.

In both examples it is to be assumed that the cooperation partners will first of all have to try out and evaluate alternative mechanisms before they are able to negotiate the optimal steering structure for their project. The more aware the actors are that the steering structure cannot be planned down to the last detail a priori, but will emerge as the cooperation system develops, the more the steering structure is likely to succeed.   ●

Success Factor – Processes

Motto: Design processes for social innovation

What have the cities of Cottbus in Germany, Saint-Denis in France, Christchurch in New Zealand and Rosario in Argentina got in common? For a number of years citizens there have been able to participate directly in determining their municipal budget. They decide for instance which infrastructure measures are to receive priority, and review at the end whether these have also been implemented. Local authorities are not legally obliged to directly involve the population in budgeting issues. Nevertheless, local policymakers and senior administrators have decided to give local cooperation systems a fresh boost by introducing participatory budgeting. This means that a new social practice has arisen in these cities, which is an example of how steering tasks for local development can be performed jointly by public, civil society and private actors.

This innovative approach first saw the light of day in Porto Alegre in Brazil in 1989. The new city government of the then recently formed labour party (Partido dos Trabalhadores) was backing social change. It invited the population to participate in planning processes at neighbourhood level. Citizens were to become protagonists of local development. Over the years the system of representation, debate, planning, decision-making, monitoring and accountability has become more and more diverse and sophisticated. It now affects many tasks of local government, as a result of which more and more processes are being adapted. The model is a success story, and is inspiring other municipalities in Brazil, Latin America and many other places in the world.

Social innovation and societal change

As in the example of participatory budgeting, Capacity WORKS sees innovation as part of social change. Developments of this kind are usually not linear. No matter how much they are planned, they cannot be organised comprehensively, nor can all decisions be taken explicitly. In many cases these developments occur without any planning at all, and change a social system through a process that is highly fragmentary and full of imponderables. The end result is a new form of steering and realisation of societal tasks. In other words, social systems possess the ability to regenerate and stabilise.

When innovations are consciously managed, the actors within a cooperation system decide to jointly master a particular challenge. They agree to drive this development through new forms of cooperation. The actors begin by describing and evaluating practices to date. This means they focus on the established processes.

Processes describe the work packages that are necessary in order to generate specific outputs in a sector. Responsibility for these work packages is assigned to specific actors. The municipalities for instance determine the need for local infrastructure projects, plan them, secure the funding and implement the projects. These processes take place on a routine basis, and are adapted to new demands whenever the responsible actors have established appropriate learning loops. One such adaptation would be involving the local population in the planning process in order to ascertain the need for local infrastructure projects.

In most cases the processes of an organisation complement those of other actors in the area of social concern. This means we can also see processes in a sector as the 'nervous system' linking up organisations. The system ensures that the outputs for which the actors are responsible are generated.

Once the actors have visualised and evaluated the established processes, the next step is to jointly formulate a change objective and agree on a path leading to the intended innovation. To determine capacity development needs, the actors also need to answer the question as to which new competencies and capacities individuals and organisations need to acquire. The complexity of this task means that no single actor can cope with it on their own, nor would any single actor possess the mandate required when operating in a context of cooperation. The lone genius making a pathbreaking discovery in the laboratory or seated at the desk is condemned to failure, because the potential innovation would not be accepted by society.

Enabling environments for social innovation

Any social innovation requires certain conditions in order to become established. Several cooperation partners must decide to replace familiar approaches with new ones. Whatever the circumstances, innovations must always be placed in the specific social context. Cultural and historical factors may create an environment within a society in which there is both a need and a willingness to experiment with new forms in order to generate joint outputs. Many innovations, however, emerge from established practices before gradually becoming established themselves, step by step.

Where innovation and social development are concerned, social systems are subject to the phenomenon of path dependency. Path dependency means that the past always continues to exert an influence, regardless of how much scope existing traditions allow for possible innovation. In this connection it is important not to see the phenomenon of path dependency exclusively as a factor hostile to innovation. Traditions also supply a society with a shared identity. They perform a stabilising role and provide actors with rules of behaviour that create behavioural certainty. Without this certainty results-oriented cooperation would not be possible, because it is this that makes the behaviour of actors predictable and dependable. Social innovations call this certainty into question, because they inevitably lead to changes in structures, processes, routines and rules. The behaviour of actors changes. Following a phase of restructuring, the cooperation system must stabilise again.

Against this background it becomes clear that the willingness to innovate is dependent on a number of conditions:

- The actors involved must be incentivised to mainstream social innovation through change processes. These processes are never wholly predictable, and therefore entail risk.

- The intended changes must be compatible to a certain degree with the prevailing value system (an important element of path dependency).

- The planned change process should be so clear that the actors involved can recognise both the possible benefits and the risks associated with an attempted innovation.

- The proposed change should first be tested in a sub-segment of the cooperation system. This would allow approaches to be modified and the process reversed at low (political) cost.

- The interim results of the change process should be visible, so that costs and benefits can be compared. Furthermore, rapid visibility ensures that the actors involved have an incentive to continue and redouble their support of the change process.

In other words, if social innovation is to succeed strict conditions must be met, particularly when we consider that the various actors within a cooperation system must agree on objectives and the implementation of specific steps. This creates a need to continuously analyse the absorptive capacity of the cooperation system. How profound can the changes be, and how much will the system take?

The specific perspective of the success factor 'processes'

Changes within a cooperation system require rules, structures, processes and rituals to be modified. The specific contribution made by the success factor (SF) processes is its focus on the relevant processes within the cooperation system and how these are linked to each other. The understanding of processes in Capacity WORKS is based on fundamental approaches to process management such as Total Quality Management, the value chain approach and Six Sigma. As outlined briefly at the beginning, this success factor focuses on the processes in cooperation systems for delivering services that are relevant to a society. To achieve this, many processes which are used to transport, read and modify information must be harmonised so that the actors involved can take decisions and implement specific activities. The SF processes provide an overarching perspective from which to visualise how a cooperation system works. The processes correspond to the key work packages. Processes can be clearly distinguished from each other. They are the responsibility of specific actors, who ideally will possess a mandate, as well as the necessary resources and capacities.

To achieve a full overview of the work packages, the actors need to establish which processes exist in the permanent cooperation system. All processes in a sector will have emerged through time, and therefore have a history. This is why we usually see overlaps, or processes that have lost significance over the course of time. When evaluating specific processes what counts most is how they contribute toward the achievement of objectives in the sector. To understand cooperation systems, it is important to focus particularly on those processes that support joint learning and cooperation by the actors involved. These processes provide information on the ability of the system to innovate.

One of the key elements of the success factor 'processes' is the so-called process map, which provides a visual overview of a cooperation system. The processes of a cooperation system are categorised according to different process types. They are built on the outputs that various actors generate. Directly or indirectly, all processes in the sector (i.e. in the permanent cooperation system) help ensure that users benefit from an output, such as access to educational or health facilities. In the same way, though, the process map may also relate to a project (i.e. a temporary cooperation system). In this case the processes identified contribute directly or indirectly toward achieving the agreed change objective.

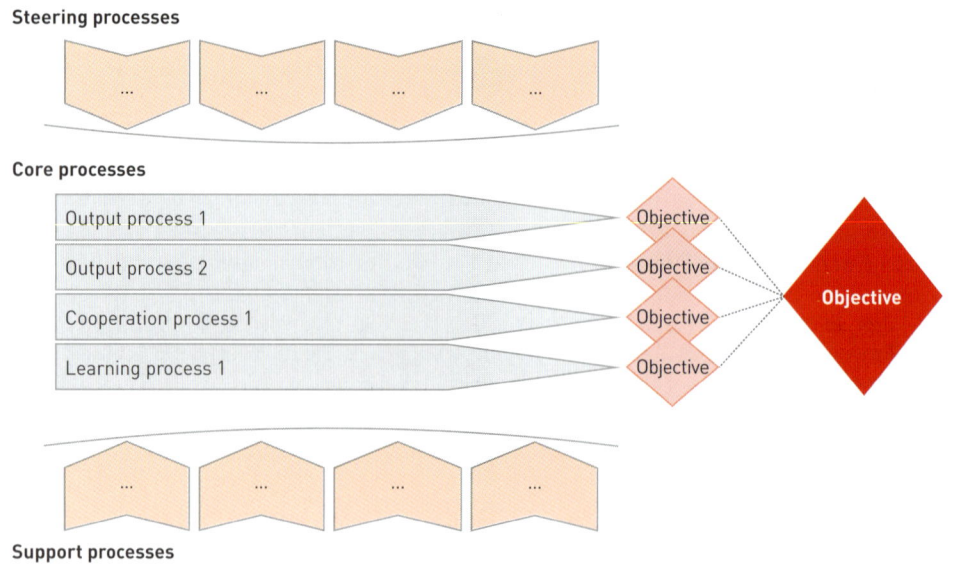

Figure 14: The process map

The starting point for the process map is always the objectives that are to be achieved. We will now illustrate this, taking access of the population to high quality health services as an example.

Every society is required to deliver high-quality and affordable health care services to its citizens. To achieve this objective, many organisations must coordinate their activities and provide their respective contributions to the cooperation system. These include the ministry of health, health insurance providers, hospital and medical associations, the pharmaceutical industry and pharmacy associations etc. Legal frameworks and appropriate funding arrangements must also be put in place. Only when all these organisations work together well will the population be supplied with appropriate health care services.

Health is a sensitive topic and one that is of major concern to the population. State and private sector actors therefore face considerable pressure to remain accountable for their actions. They can only meet citizens' expectations if the processes for which they are responsible actually help achieve the objectives.

From this fragmentary description it is already evident that a number of processes are needed within the cooperation system. These different processes must be mutually harmonised and efficiently managed. The interplay between them should be such as to ensure that each actor involved is able to make their contribution in a way that enables the system to deliver good results at reasonable cost.

To analyse the status quo of a cooperation system and identify the need to change it makes sense to distinguish between specific types of processes.

The *output processes* are the processes that relate directly to the objectives of the cooperation system. Key elements in supplying the population with health care services include the treatment of patients in hospitals, and the services provided by general practitioners and home care providers.

The *cooperation processes* support the output processes by coordinating the various actors. For example, the health insurance companies and service providers (such as hospitals, doctors or pharmacists) cooperate with each other in order to facilitate payment for the services provided.

The *learning processes* are needed because they enable the actors to appraise the quality of service delivery in the sector and make needed changes. Hospitals, research institutions and health insurance companies for example monitor the therapeutic results of new treatment methods, and offer these to patients through new output processes.

Output, cooperation and learning processes are closely intertwined and have a direct effect on quality within the cooperation system. This is why they are referred to collectively as **core processes**.

The **support processes** are work packages that underpin the other types of process. They do not play a direct part in generating outputs, however. For example, nurses, pharmacists and health insurance professionals are trained in order to ensure that the health sector has access to a pool of properly trained young professionals. This maintains the long-term availability of the output processes.

The **steering processes** are the ones that set the legal, political and strategic framework for the other types of process. They supply the cooperation system with decisions. In our example this would include setting the compulsory insurance premiums, selecting the services provided to insured individuals and determining the costs reimbursed to the service providers.

Situations can arise in which there are no clearly defined objectives for the cooperation system, or in which several objectives exist that conflict with each other. This information is highly relevant when seeking to understand a cooperation system, and provides pointers as to why specific processes are proving ineffective. The actors involved are then able to identify realistic paths for change and innovation. To achieve this, however, it is first of all necessary to gain additional perspectives on the cooperation system. The analysis of actors and their interests provides valuable information on existing potential for cooperation, as well as lines of conflict. This supplements the perspective on the processes of the system.

The ultimate results of a project occur in the permanent cooperation system. These results are attributable to innovations. Innovations usually affect many processes, and help reorganise the interfaces between them. The sector thus adapts to new demands. In other words, this means that learning takes place on all levels of capacity development. Societal frameworks change, as do the forms of cooperation between the actors. Organisations must restructure the range of services they offer, and their staff members require new competencies.

The process map provides a strategic perspective on the sector. The operational planning activities designed to change selected processes can be supported through the so-called process hierarchy.

The process hierarchy is used to analyse the selected process in detail by visualising its sub-processes. These can be broken down in turn into their constituent elements like Russian dolls, down to highly detailed actions of individuals contributing to the process as a whole. The degree of detail needed will always depend on the requirements of the specific case.

The challenge: first understand and then evaluate the permanent cooperation system

When working with the SF processes it can be tempting to construct a desirable picture of how the current situation in the area of social concern 'should' look. However, to identify potential social innovations the cooperation partners must exchange views on the 'realities' they each see, and negotiate a joint picture of the existing objectives and established processes in the sector – even if these are sometimes unclear or do not 'work' properly. This approach enables the actors to reach a shared understanding of what they regard as processes in the sector that already work well. Only then will it be possible for the actors involved to determine where processes are either lacking, not mutually harmonised or working badly, and why this is the case. The issue of what need for change exists and what reforms it might be appropriate to initiate should not be addressed until this analysis has been completed.

This approach will also help avoid normative ways of thinking on how a particular area of social concern 'should work'. A solution that had already been defined at the outset would usually lead to too strong a focus on the deficits in the existing cooperation system. Solutions to societal issues are highly specific and usually cannot be transferred from one context to another.

It is easier to access the existing cooperation system on its own terms by pursuing an initial approach that is descriptive and does not make any premature value judgements. This is key to identifying the energy for change among the actors in the sector, and using this as a resource for successfully mainstreaming social innovation. A project and the permanent cooperation system can then be matched so that the desired change can succeed.

The actors who will ultimately also mainstream the innovations in the system should always be involved in this work. The SF processes helps the different actors agree on a joint perspective. It also helps them supplement this with a SWOT analysis (SWOT is an acronym for Strengths, Weaknesses, Opportunities, Threats) and feed it into a broader strategy-building process.

Practical consequences for using the success factor 'processes'

With its analytical incisiveness and its focus on concrete processes in the sector in which social innovation is to take place, the SF processes complements work with the other success factors of the management model Capacity WORKS at many points.

- Sustainable changes do not occur automatically when new rules are adopted and organisational and cooperation structures are modified. The quality of outputs will only change if processes do likewise.

- Focusing on objectives in the sector supports the results-orientation of the actors involved. The objectives relate to people's access to quality services and public goods at appropriate cost. Each project must ensure that its orientation matches the changes in the sector which the actors have agreed on.

- Focusing on processes in the sector helps identify strategic options for the project, because it locates the processes that might create leverage.

- Distinguishing between types of process within the sector helps clarify the mandate of a project. Improving specific support or learning processes requires a completely different kind of political will than the one required to change steering processes. The latter ultimately involves politically sensitive issues, for instance when budgets need to be allocated or responsibilities defined. As a consequence a project's steering structure must be organised such that relevant actors in the sector get involved and jointly shape processes for social innovation.

Internal processes within the project

Besides describing the area of social concern the SF processes can be used to analyse the **internal processes in the project**. These internal processes are designed to initiate the intended innovations in the sector. The example below shows how the SF processes is applied to the internal management of a project.

Process map for a project on 'Introducing participatory local budgeting'.

Participatory budgeting enables citizens to have a say in how funds are spent in their municipality. An analysis of the permanent cooperation system 'local government services' shows that the relevant actors are interested in this innovation. The association of local authorities is willing to solicit the support of mayors and local councillors for this innovation. It hopes this will give a boost to the country's decentralisation policy. Citizens' associations and NGOs are interested in direct political participation and accountability because this will further democratise their country. The ministry of the interior supports the idea because it believes this will boost the modernisation of public administration. A project is then established.

For the project strategy an option is selected that involves gathering experiences in at least 10 pilot municipalities. All parties concerned share the hypothesis that the maximum impact will be achieved by integrating participatory budgeting into the annual municipal investment planning process.

To gather experiences in the pilot municipalities, the project needs at least two output processes. First of all the political decision-makers must be sensitised so that they can agree to changes in the way the budget is adopted. Secondly the budget planning processes for local investment must be professionally moderated, which means making the necessary expertise available.

The success of the piloted experiences will require close coordination and harmonisation between political decision-makers, representatives of the municipal administration and the population. This will require a cooperation process based on clear expectations on the part of local NGOs concerning the amount of investment funding available.

The results then need to be jointly evaluated by the actors involved, and lessons learned from this. This is the only way to transfer the initial experiences of the pilot municipalities into new processes, and make a contribution to the national political debate on decentralisation.

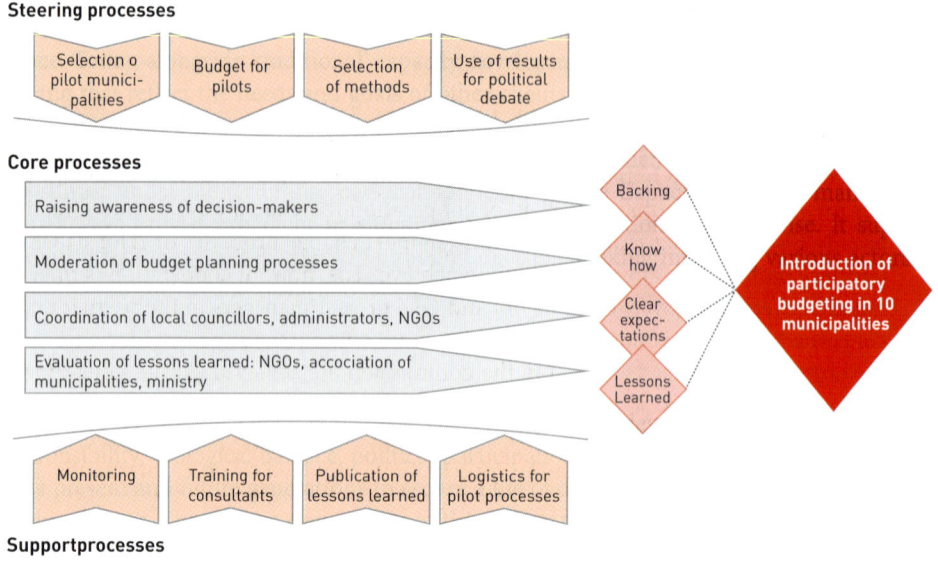

Steering processes

| Selection o pilot munici- palities | Budget for pilots | Selection of methods | Use of results for political debate |

Core processes

Raising awareness of decision-makers

Moderation of budget planning processes

Coordination of local councillors, administrators, NGOs

Evaluation of lessons learned: NGOs, accociation of municipalities, ministry

Backing

Know how

Clear expec- tations

Lessons Learned

Introduction of participatory budgeting in 10 municipalities

| Monitoring | Training for consultants | Publication of lessons learned | Logistics for pilot processes |

Supportprocesses

Figure 15: Process map for 'Introducing participatory budgeting'

Implementing the core processes of the project requires a number of additional steering and support processes to guide the cooperation partners. For the steering processes, appropriate pilot municipalities must be selected as cooperation partners. Budgets must be made available so that the pilot lessons can be learned. Furthermore, the right methods have to be selected to implement specific activities. Decisions must be taken regarding the form in which the results are to be fed into the policy debate on decentralisation.

Project progress is tracked by a monitoring system. This is an important support process that underpins possible changes in steering decisions (e.g. selection of different methods). To guarantee the quality of consulting services, training must be provided to ensure that the process moderators possess the necessary expertise. The logistics of these processes must also be ensured, for instance regarding the provision of appropriate premises, invitation management and necessary working materials. Publication of the lessons learned is a further support process designed to disseminate these results beyond the pilot municipalities.

If the project needs to analyse certain processes in more detail for purposes of operational planning or monitoring, one approach would be to prepare a process hierarchy for the output process 'Raising awareness of decision-makers'. Within this output process a distinction is drawn between training, field trips and coaching. If the sub-process 'field trips' then need to be analysed in yet more depth, this can again be broken down into smaller units. One of these elements, the selection of appropriate cases, could if required be broken down into its own constituent elements.

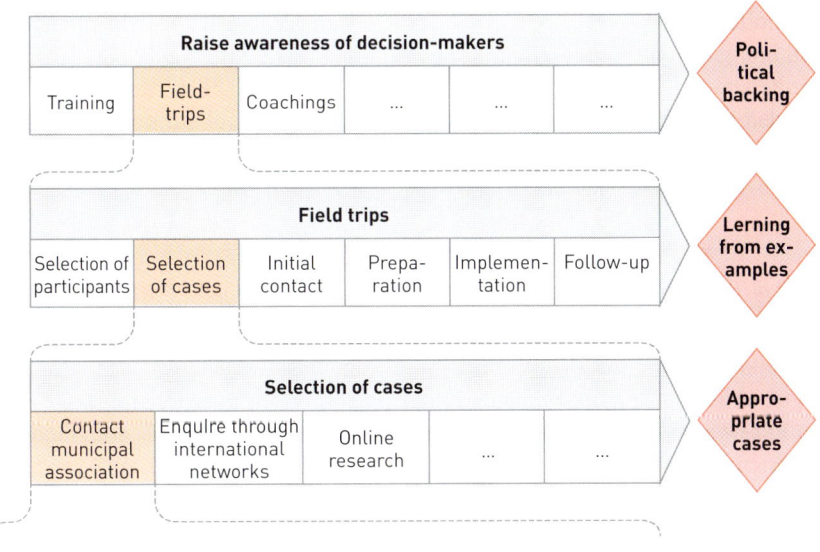

Figure 16: Process hierarchy for a selected core process

Quality criteria for processes

Processes in cooperation systems are sensitive and liable to malfunction. This is why we strongly recommend that you document the joint core, steering and support processes in detail, and discuss them explicitly and openly. Who will do what, by when, who will take the next step, who will be responsible for the process (sub-process)? The processes within the project should meet the following quality criteria, so that they can help form models within the sector:

- Each process must make a clear contribution to the results of the project.

- The processes should be (increasingly) stable and not require continuous reorganisation.

- The speed of the processes should be appropriate in relation to the dynamics of the cooperation partners involved. A quick process is not always a good process.

- The processes should be time-efficient and cost-effective.

■ In-process learning and optimisation should be discussed during cooperation between the actors involved, in order to identify problems and bottlenecks, as well as potential improvements, and make adjustments.

The cooperation system should make sure that all actors involved assume responsibility for the processes within the project. It must also organise processes to continuously link up the contributions made by the organisations involved, and the reality of the sector in which the intended social innovation should ultimately become established.

The concept of the results model was presented and discussed in the chapter on objectives and results. The point or points at which a project is to generate results can always be located using the process map for the sector. Which processes in the permanent cooperation system are to be modified? How will these changes affect other processes, for instance if the modification of a steering process leads to a reallocation of mandates? Will the responsible organisations then be able to better organise their output processes? Will it then be possible to better achieve the objectives in the sector? Will citizens gain improved access to social services? The results model makes the results hypotheses underlying the change explicit, while the process map shows exactly where in the system the results will occur.

Success Factor – Learning and Innovation

Motto: Focus on learning capacity

How do cooperation systems learn? And how can we tell that they have learnt something? The answer is, when a cooperation system adapts to changed requirements. Successful cooperation management focuses on ensuring that learning capacity is developed on all three levels. Within a society, frameworks are adjusted, and cooperation relationships improve. Organisations then learn to continuously raise quality as they help achieve the joint objective. And the individuals within those organisations develop their competencies, and jointly shape learning processes so that they can help generate sustainable results in their specific context. In conjunction with the changes within organisations and cooperation systems, these human learning processes help create an enabling environment for launching and operationalising innovations.

When they hear the term 'learning', many people first of all associate it with individual human beings. We can see individuals and interact with them. We can study their learning processes by directly surveying and observing them.Libraries are always well stocked with books on human learning, arranged by criteria of age or life situation of the learners, or learning content. Learning supports the cognitive, emotional and social development of people. Our knowledge on how people learn is immense. But what do we know about how organisations learn? Do only the right members of the organisation need to learn the right things in order for the organisation as a whole to function better?

If an organisation is to change the way it works, then the learning processes needed appear to be far more complex than the learning processes of individuals. Joint action by individuals within an organisation is strongly influenced by the rules, structures, processes and rituals according to which the organisation operates. These must first of all change in order for the whole organisation to learn, and hence change. Much less is known about how organisations learn than about how individuals do.

In cooperation systems comprising several organisations, the issue of learning becomes an even bigger challenge. It is not possible to survey or interview cooperation systems 'per se'. We can only interview individuals who represent the organisations that make up the cooperation system. As in single organisations, the system only changes when rules, structures, processes and rituals undergo changes. These changes must be negotiated between the actors involved, and require a joint will in order to become sustainable.

Like individuals, organisations and cooperation systems generate action. In so doing, they cannot avoid learning. When organisations and cooperation systems act they always learn. The relevant questions are 'what' they learn and 'how' they learn it, because the 'what' and the 'how' determine the usefulness of what they have learned.

Without entering into the questions of what should be learned and how, we note that any learning in the first instance remains a learning of lessons on a 'one-off' basis. So, what is required is a learning capacity that will enable individuals, organisations and cooperation systems (if not to say the society as a whole) to cope successfully with future challenges. In order to grasp what is needed for a change for the better in an area of social concern, we should focus our attention on capacities for learning on the various levels right from the start. And we should be looking at both issues, i. e. what kind of changes would help on the level of individuals, organisations, cooperation systems and framework conditions, and how learning capacity can be fostered on each level more generally. This is what the success factor 'learning and innovation' is all about.

Learning how to learn

If a cooperation system could speak it might tell the following story:

'If you want to understand me, then you must begin by asking what I'm there for in the first place. I'm supposed to ensure that small and medium-sized enterprises (SMEs) in our region are able to develop so that as many people as possible get a job. Long before I came along there was no need for me, because the SMEs in the textile, leather and mechanical engineering sectors had a long tradition of selling their products for good prices on the national market. But then the crisis began. Suddenly cheaper products from other countries started coming in. For a long time our companies tried to produce more cheaply themselves. But this also led to a decline in quality standards, and fiercer and fiercer competition with the imports. More and more enterprises collapsed, and unemployment grew.

It's difficult to say exactly who it was who got the ball rolling to create me. The municipalities were alarmed, and wondered what they could do in order to strengthen the weakened enterprises. The public administration granted trading licences and building permits, raised fees and taxes, but did not do anything else for the local economy. People from the local chambers of trades got together and tried to find a way out of the crisis. Up until then they had not had many dealings with each other, because the various chambers only took care of their own members.

At some point it became clear to all the actors involved that in order to make progress they would be dependent on each other. The first meetings were held in order to agree on a contingency plan. How could SMEs be provided with improved access to affordable loans? How could SMEs integrate new technologies as quickly as possible in order to reduce costs? How could people who had already lost their jobs be supported in order to prevent them from leaving the region?

When the people involved got round to agreeing on which issues they could tackle quickly, and who would make what contributions, a hard core of actors emerged. After that, new potential partners were approached: credit cooperatives, local banks and the regional government. And all of a sudden there I was – a cooperation system – because the actors had a joint objective. They made their contributions to me, carried out first activities and began planning on a regular basis. Since then they have called me 'regional cluster'. Over the last 20 years, I have gradually grown to include new actors, new areas of social concern and new processes. My most recent achievement is enabling universities and SMEs to work together on developing clean technologies. There are now even one or two SMEs that export this kind of technology.

Over time I have grown, and I'm constantly changing. Are the actors who form me aware of that? Do they plan my development, for instance? Well, every now and again some individuals within the administration say they can control me completely. Other actors from the private sector say the same thing. As a result, the two cancel each other out. Nevertheless I carry on developing, because I possess the 'mysterious' ability to steer the direction in which I go. Which is a good thing, because the cooperation partners don't actually know how they do it. Particularly at the beginning they underestimated my capacity to learn, and some actually believed that the capacity of the individuals involved would be the crucial factor. This is presumably the reason why they first of all implemented action plan after action plan, without asking themselves at an early stage how my change process could be managed. Time and time again, though, the results this produced provided the necessary feedback.

A short while ago, several particularly committed actors joined forces within my boundaries. They want to understand me better, so that they can more consciously influence the necessary learning processes. They concede that they will never be able to fully predict or even control my behaviour. They do not understand absolutely everything about me, but experience has taught them a few lessons:

- Although the individuals involved are willing to learn new things, the organisations are often reluctant. At first glance this seems paradoxical. For example, it was clear to most of the SME proprietors that the enterprises had to change. Yet these enterprises had grown up slowly and successfully over generations. Changing too radically and rapidly would disrupt everything. Nevertheless, production operations would have to change, and products sold to new customers whom the enterprises did not yet know etc. If they did not adapt, companies would collapse. Yet adapting continuously without returning to a rhythm based on routine would also lead to collapse. So companies had to strike the right balance between stability and change.

- It was difficult to face up to the fact that, in order to survive, we needed to abandon and let go of tried and tested ways of doing things. After all, everyone was proud of the success stories of the past and was saddened by the loss. The actors needed a perspective within which the new could take shape. Others had already undergone similar experiences before us. When the actors drew lessons from these experiences, and found that not everything from the past was bad, they also became willing to try out new things.

- The enterprises in particular felt very safe in the knowledge that the aim was not to learn as many new things as possible at all costs. Since the aim was to become competitive by also highlighting the unmistakable features of our region and its products – our unique selling points – the actors decided not to change certain things.

- As there were strong feelings of insecurity, particularly at the beginning, every setback caused heated debate. Many times the actors wanted to do away with me as a cooperation system and go their own separate ways. Yet increasingly they came to see that they were dependent on each other, and somehow they began to see setbacks as signs that they needed to continue learning.

Change takes time. Learning processes cannot be speeded up at random. Individuals learn more quickly than organisations, and organisations learn more quickly than me, because cooperation systems include many variables that no one fully comprehends. At some point the actors understood this. Learning takes place on many levels that interact with and influence each other. Not everything can be changed at the same time. That's why I need a learning architecture to provide a clear picture of how things build on and interact with each other. And since no one can predict precisely how I will react, the actors need to try things out, observe what happens and continually adjust the learning architecture. This is how the actors have learned to learn.'   ●

Learning as evolution

This example clearly shows that organisations and cooperation systems develop, adapt to their environment and even influence it. This is only possible when learning takes place within the cooperation system. Changes within organisations and cooperation systems are innovations, because they involve achieving things in new ways. One example of new patterns and routines emerging in cooperation is when businesses clearly communicate to universities what their practical education and training needs are. Universities respond to this by establishing new internal processes in order to deliver the services for which there is a demand. Although these learning processes cannot be planned exhaustively, a structured approach will make them more likely to succeed.

We can explain how learning and innovation take place within organisations and cooperation systems using the three basic mechanisms of evolutionary theory.[10]

In organisations and in cooperation systems, minor or major deviations from the established routine emerge at various points either as a result of planning or spontaneously (**variation**). These variations are often prompted at the interfaces between organisations and their environment, or between cooperation partners. For example, a client might express a special request that cannot be met using existing standards. Or the users of public services might draw the attention of state actors to quality problems. Within an organisation, staff might develop ideas for designing new processes or services. Some organisations and cooperation systems develop numerous variations, while others lack variation.

One way of transferring lessons learned on innovation across cooperation systems (and beyond them) is knowledge sharing. For example, representatives of organisations from different societies might jointly seek variations to help them deal successfully with specific challenges. Some participants may already have some relevant experience. They then share their knowledge by presenting the lessons they have learned in a way that enables the other participants to translate and transfer this into their own setting.

Too many variations can leave staff or involved actors feeling uncertain as to how they should act in the face of such diversity. Whatever the tendency/organisational pattern is, i.e. whether variation is non-existent, infrequent or abundant, if it is to become meaningful for the whole system then it needs to be acknowledged and approved. At a certain point line management or a steering committee will take a decision and select from among the known variations the one that seems the most suitable. For this step (**selection**), it is key to organise the relevant decision-making processes appropriately. Organisations and cooperation systems differ in terms of their decision-making processes. Some support swift selection from among the available variations, while others involve lengthy processes. However, in the end the decision reached will have the backing of the relevant actors, and can be operationalised.

Once a selection has been made, this still does not guarantee that the variation will exert influence on the system. Interventions are required to stabilise the innovation within the system (**stabili-sation** or **re-stabilisation**). This means that rules, structures, processes and rituals are reviewed, and where necessary adjusted. New routines then emerge. This is how the cooperation system or the organisation gains the stability it needs in order to survive. These interventions require a great deal of attention in order to ensure that the actors in the organisations or cooperation systems concerned can rapidly rebuild confidence and certainty in the way they act. Since each system has its own specific learning capacity and patterns, each one will deal differently with change, conti-nuity, and the balance between the two. Some master change well, while others fail in adjusting to their changing contexts. The capacity for managing change processes is crucial and needs to be worked on.

Organisations and cooperation systems thus learn by means of variation, selection and stabilisa-tion/restabilisation – and successful projects support this learning mechanism by initiating tai-lor-made interventions. In some cases, structures and processes are established in order to gen-erate variation. Decision-making processes can also be improved in order to boost selection. It is also conceivable that situations will arise in which change management as a whole is made the focus of the interventions initiated. Learning by organisations and cooperation systems is always geared to their objectives. The results delivered by the new patterns are manifested in benefits for customers, clients, stakeholders, shareholders and other actors.

Sustainable learning on the levels of capacity development

Although the concept of capacity development originally came from international cooperation, it can be applied to any kind of cooperation system. 'Capacity' means the ability of people, organisa-tions and societies to manage their own sustainable development processes and adapt to changing circumstances and frameworks. Developing it also means strengthening the proactive management capacity of the actors within a cooperation system. Change processes are operationalised in such a way that the innovations they generate are sustainably mainstreamed through new routines.

We distinguish between the different levels of capacity development that have already been men-tioned several times: (1) societal and policy frameworks, and the cooperation relationships be-tween the actors involved, (2) organisations, and (3) individuals. The proactive management ca-pacity of a cooperation system, and thus its effectiveness, are determined not only by specific capacities on the different levels, but also especially by the interplay between them.

When facilitating learning in a cooperation system we need to take these interactions into account and use them. The relevant actors should aim to jointly manage change and take effective action to bring about sustainable development. A key role is played in this context by the actors learning both with and from each other, and thereby innovating.

Capacity development is fundamentally a process which must be driven by the actors themselves. This presupposes a high level of ownership, i.e. identification and commitment, on the part of those involved in achieving the intended changes. This ownership will usually emerge and grow in the course of a change process.

External actors often support the capacity development process of cooperation systems by delivering consulting services. These consultants act as catalysts, because they do not pursue any interests of their own, and the other cooperation partners trust them. On the one hand this provides an enormous opportunity because of the incoming resources. On the other hand, if the actors in the cooperation system do not shoulder their own responsibility, but try to give it away to these external consultants, this always creates a risk. This is so especially since at the same time it is tempting for consultants not to confine themselves to their limited role but to take on the managerial responsibility offered to them. Unfortunately, this jeopardises the sustainability of results. This is why in cooperation systems which are supported by external actors such as consultants we need to make absolutely sure that the energy for change is coming from the actors involved themselves.

Learning at the level of society – policy field and frameworks

Any cooperation system operates within a policy space that sets key frameworks. Often these can only be changed with very great difficulty. The frameworks are determined by a fabric of rules, processes, structures, relationships, organisations and individuals. Cooperation systems are embedded within these frameworks, and at the same time try to influence them. To manage cooperation systems successfully it is important to understand how these elements interact. First of all we need to obtain a clearer picture of which objectives can actually be achieved. Secondly we need to generate important conclusions concerning how to manage the cooperation system, for instance with regard to which actors should be involved and what the appropriate forms of cooperation would be.

Here are some of the key elements that determine the frameworks within which a cooperation system operates:

- the traditional and cultural context of a society, which is manifested in values and social patterns (basic orientations with regard to the role of the state, structure of the economy, status of specific population groups within a society etc.)
- the history of the relevant institutions and organisations
- the system of incentives, and fundamental beliefs of the relevant actors
- the strategies, objectives and interests of the key actors
- the actual decision-making routines of the relevant institutions and actors
- laws, rules of implementation and quality standards for public services.

These and other frameworks change when learning takes place at the level of the society. The term **scaling-up** refers to a consciously selected and targeted way of disseminating and mainstreaming lessons learned and knowledge across a policy field. The process is designed to sustainably scale-up innovations with structure-building effects. We can distinguish between three types of scaling-up:

Vertical scaling-up involves institutionalising approaches that have already been successfully piloted, for instance in a project. These piloted approaches are usually institutionalised at the national level in the form of laws, policies, national development plans and programmes. This is how innovative lessons learned are sustainably disseminated across the board within the permanent cooperation system.

Horizontal scaling-up involves transferring approaches that have been successfully piloted directly between similar organisations, e.g. universities adopt new curricula, municipalities adapt their administrative procedures or small farmers' marketing cooperatives copy a successful commercialisation strategy. Unlike vertical scaling-up, horizontal scaling-up does not need national standards or policies to be adjusted. The results of the project are scaled-up through the processes of learning and cooperation within the permanent cooperation system.

Functional scaling-up involves transferring strategies, approaches, methods, and lessons learned, etc into a new context.

Learning at the level of society – cooperation relationships

Cooperation systems are excellent learning arenas because they are an environment in which the actors involved seek joint solutions. Sustainable change is based on the fact that the cooperation partners develop their resources and capacities together. Within the cooperation system, the objectives can only be achieved when the actors learn jointly. In these settings learning processes can go beyond the boundaries of a policy field.

Cooperation systems are organised so that several cooperation partners can work together as effectively as possible in order to achieve benefits in a specific societal context that cannot be achieved by any of the cooperation partners on their own. When cooperation systems learn and social innovations become established routines, elements change whose specific features make the cooperation system concerned unique. Returning to the example of SME development described above, this might look as follows:

- **Structures** are formed that result from the objective of SME development, the policy fields of access to credit services and technology development, the private and public sector actors involved, and the specific forms of cooperation.

- **Processes** evolve that gear contributions made by the partners involved to the objectives of the cooperation system. For example, chambers and public administrations provide technological advisory services on industrial estates to support enterprises in proposing high quality investment projects to local banks and credit cooperatives.

- **Rules** for cooperation and decision-making are newly established in the cooperation system, for instance regarding the adoption of action plans, or the selection of university research projects that are eligible for support etc.

- **Rituals** originate within cooperation and form an integral component of the culture of the system. They are of high symbolic value and create identity for the actors involved. They include for instance the logo of the cooperation system, which taps into the glorious tradition of the old trade guilds, as well as specific festivals that celebrate successes or appeal to the actors' sense of cohesion.

Cooperation systems are characterised by a rich diversity of ideas and actors, which makes them especially suited to designing and managing learning processes in modern ways. Successful cooperation systems learn in two respects. First of all structures, processes, rules and rituals are adapted according to the current prevailing objectives. Secondly, though, the actors also learn how they can respond in similar situations. In other words, they learn to learn. This creates a setting in which new initiatives for further innovation may emerge at many points within a cooperation system.

Since the cooperation partners must reach agreement as to whether and how they will initiate these change processes, resources are tied up time and time again. Cost-benefit analyses are also crucial for learning. This makes it necessary to consider and document whether the project objectives and the learning goals have been achieved. The actors must define and monitor learning objectives and indicators just as much as they define and monitor the objectives resulting from the joint strategy, such as establishing a new credit line for technology development. In projects, actions are tested in a multi-organisational context with a view to determining their sustainability. In this sense the whole project is nothing but a model for learning by the permanent cooperation system.

Learning at the level of organisations

Organisations are 'entities' in their own right. The widespread belief that organisations can be reduced to the number of their members falls short of the mark. Organisations generate their own meaning – and have their own rationale. They usually develop in response to issues of concern within a society, and supply corresponding solutions. They succeed sustainably when they provide solutions for tasks at transaction costs which are lower than they would be without the organisations concerned. In the course of their history organisations develop an autonomy that is manifested in specific structures, processes, rules and rituals. Organisations are agglomerations of objectives that emerge through results-orientation and formal membership. These objectives define the framework for the organisation's operations.

When they become members of a cooperation system, organisations are repeatedly confronted by demands placed on them from within the cooperation system. Organisations need to cooperate with each other in order to achieve objectives that no organisation can achieve on its own. This creates fresh opportunities and potential. However, the organisations also have to keep adapting in response to the demands placed on them. This is why the individual partners within a cooperation system should develop the capability to absorb stimulus from within the cooperation system, continue developing their capacities and supply the inputs and contributions expected of them.

One way organisations achieve this is by developing new structures, rules, processes and rituals that cover the area of cooperation with other actors. This creates scope for action and decision-making that enables individuals within the organisations to devote a portion of their energy, working time and resources to cooperating with others. It may also be necessary for these individuals to develop their competencies such that they are able to make exhaustive use of this scope, and increase it, in order to represent their organisations appropriately within the cooperation system.

Organisations learn by incorporating experiences and knowledge into their structures, processes, rules and rituals, and making these intelligent. Organisations learn for instance when

- a restructuring process provides them with new experiences in designing and changing organisational structures;

- value creation processes are adjusted, thus optimising the organisation's business as a whole;

- systems of rules – such as systems of rules for human resources development – are changed;

- quality management systems are established to ensure that the standards set are complied with when supplying products or services. Organisational learning also includes establishing capacities to continuously identify and meet learning and improvement needs.

Learning at the level of individuals

Learning and innovation are often initiated by individuals who either see new opportunities and potential, or simply see an imbalance between the way things are supposed to be and the way they are in reality.

Developing capacity at this level means strengthening the competencies of human individuals. Yet this does not take place in a vacuum, independently of relationships, links and context. Activities at the level of individuals develop their full potential when they are oriented toward the system of reference, and when the links to the other levels of capacity development are also effectively managed at the same time. What we mean by the 'system of reference' is the sphere of direct influence by an individual – which may be an organisation, a cooperation system, a network or an informal community. Developing an individual's competency helps enable that individual to initiate and facilitate change in their respective systems of reference. As change agents, individuals are enabled for instance to manage processes of sharing within their professional network more efficiently, to prompt reorientation in their environments, or to act as disseminators to consolidate learning in organisations, networks and policy fields.

Human Capacity Development at GIZ

GIZ possesses years of experience with Human Capacity Development (HCD). GIZ sees HCD as a service package that addresses learning by individuals as part of capacity development processes. HCD organises the learning settings of relevant members of the cooperation system in order to enlarge their respective individual contribution to change in their home organisation and the cooperation system.

Using the learning architecture, we identify various competency needs for the HCD activities in relation to the objectives of the cooperation system. We make six service pledges for our interventions at the level of individuals. These reflect the various ways in which we focus on developing HCD activities tailored to the relevant target groups:

1. Make individuals more effective

By learning strategies for self management and knowledge management, individuals develop their personal and social skills. To achieve this they also reflect on their own actions and their understanding of their roles.

2. Strengthen the proactive capacity of experts

By supporting the development of practical professional competency, we enable experts to initiate and facilitate change from their own perspective as specialists. This involves focusing on specific cases, and developing capacity to continue coping successfully with change by pursuing systematic 'continued learning'.

3. Strengthen the creative potential of managers

We familiarise individuals responsible for projects or processes (i. e. managers) with the concepts of efficient process and change management. We thus enhance their abilities to shape institutional and societal change for sustainable development self-reliantly and in the long term.

4. Strengthen trainers and advisers in their role as disseminators

Trainers and advisors hone their didactic and methodological capacities in order to maximise their effectiveness as disseminators both when supporting individual learning processes, and within organisations and cooperation systems.

5. Build and consolidate leadership responsibility

We enable leaders and change agents to continue developing their leadership and strategy-building skills. This enables them to more effectively drive comprehensive change within their own system of reference. We pursue a dialogue- and values-based approach that highlights the responsibility of leaders for the common good.

6. Network individuals for sustainable learning and change

People learn most easily in communities of learning, regardless of whether these communities are face-to-face or virtual. This makes knowledge available worldwide. As practitioners share experiences and lessons learned, this generates regular impetus for change within their respective working contexts. In many cases these networks between individuals have existed for a long time, which means they can manage themselves as they continue to find appropriate solutions to current challenges. Local, regional and global alumni networks often help make such networks for learning and sharing sustainable.

By combining these different focuses we are able to boost the development of comprehensive proactive capacity, and further increase effectiveness in the various contexts and for the various needs.

How do individuals learn sustainably?

People learn with and from each other (peer learning), and share their knowledge and experiences (knowledge sharing). This is possible thanks to the enormous capacity to learn that humans possess, which goes beyond the mere accumulation of knowledge. Human beings learn by acquiring complex experiences and interacting with other people. When lessons learned are transferred, the individuals involved evaluate and modify their existing beliefs and views. A willingness to integrate new experiences into our perception and review their validity (i.e. the capacity for self-critical thinking and reflexive learning) is a key prerequisite for change. The deeper insights we gain as a result provide us with a basis on which to review our own actions, thus creating the possibility for us to change them (i.e. to learn).

At the same time learning is always linked to capacities and values that are culturally reinforced and that define the role of the individual within society. Consequently, ideal learning conditions will take account of the cultural, social and political context. In social change processes, this is crucially important in order to ensure that the learning requirements made on the individuals involved can be contextualised appropriately. Furthermore, if learning settings include spaces where the individuals concerned undergo shared experiences, joint attitudes can emerge.

Adults learn best when their own biography is taken as a starting point, and spaces are created where they can reflect on their own experiences and knowledge, and compare these with fresh stimuli. This is how individuals identify new ways of acting. When people wish to change the way they behave this often requires enormous effort. How successful they are will depend largely on whether the individuals concerned have seen themselves as effective actors in their own context, and receive recognition and appreciation for their changed behaviour. So, within the organisation in which an individual works, or within the cooperation system, the opportunity must exist for that individual to put into practice what they have learned, and in so doing to initiate and help manage change processes.

One particularly important kind of learning is on-the-job learning. People do learn in the course of their day-to-day work and the challenge is to manage and make meaningful use of these opportunities for learning. This can be achieved for instance by giving individuals new jobs to do, involving them in project teams, inviting them to spend time in an area of work with which they are not familiar, or enabling them to get involved in cross-organisational communities of practice. This type of learning need not be managed through formal training courses. Instead it simply requires attention to be drawn to those opportunities for learning in day-to-day work, and the organisation of informal networks.

Key to effective competency development is matching up the skills that already exist with those that we would like to develop. The skills to be developed result from the specific tasks that the individuals concerned will be required to perform. The first step is to perform a skills gap analysis. This shows which skills we need to focus on, allowing us to identify corresponding learning measures.

Competency development is based on skills profiles that guide the learning process, and include various perspectives, expectations and experiences of the actors involved. Experience has shown the following five dimensions to be important: technical expertise, methodological expertise, managerial and advisory expertise, training expertise, communication and cooperation expertise, and self-management expertise.

A modern learning culture supplements the necessary face-to-face contact with further elements. Today, people no longer need to gather in one place in order to learn together. Web-based forms of learning (such as e-learning, e-coaching and e-collaboration) are becoming increasingly important. In all cases where learning is designed to change the way people act, transform their views and interpretations, or develop their emotional competence, face-to-face contact (e.g. training,

workshops or coaching) remains indispensable. This is where a modern culture of learning identifies innovative approaches by relying increasingly on forms of joint (intercultural) learning and
facilitating peer-to-peer learning.

The specific perspective of the success factor 'learning and innovation'

Since cooperation systems continuously adapt to their environment, learning processes take place
permanently, regardless of whether they occur contingently or are consciously managed. Since
projects aim to sustainably mainstream the desired changes within an area of social concern,
learning is their key theme. This is why it is essential to closely interlink any project with the relevant area of concern. The SF learning and innovation helps manage these processes. At the same
time it repeatedly warns us against believing that actors can control their cooperation system.

Projects that are successful in the context of cooperation make use of their ability to develop models on all levels of capacity development. They are arenas of learning in which numerous possible
innovations can be tried out before being mainstreamed in the area of social concern through
fresh routines.

The art is to create a 'composition' of interventions that support and reinforce each other on the
different levels. A **coherent learning architecture** of this kind ensures that people possess the
skills necessary to take on new tasks, such as those that might have become necessary due to a
redefinition of an organisation's objectives. Ultimately, however, it will not be sufficient for individuals only to extend their knowledge base, and change their attitudes and behaviours. They will
only be able to make effective use of these competencies when the structures, processes and rules
of the organisations involved and of the cooperation system as a whole change. This means that
frameworks and opportunities must be created to enable individuals to bring their new expertise
to bear, and to enable new routines to develop and become established on the various levels.

The SF learning and innovation aims to explore spaces for learning, and bring these together in
a joint strategy to strengthen the proactive management capacity of the actors involved. The SF
learning and innovation supplements all the foundational elements of the management model
Capacity WORKS by strengthening the following capacities, which can also be used outside the
project context:

- objectives- and results-oriented management, to enable the cooperation partners to self-reliantly continue orienting their management practices to newly agreed objectives and results;

- strategy development, to enable the actors to develop options that broaden their scope for
  action, and open up pathways that can give a cooperation system direction and meaning
  even in turbulent times;

- cooperation management, to enable the actors to adapt the forms of cooperation and selection of cooperation partners to the current needs at any given point;

- steering mechanisms, to develop the capacity to establish appropriate steering structures
  within the cooperation system, and modify these as required;

- process design, to enable actors to manage processes in the area of social concern and internal processes in a way that allows social innovations and learning mechanisms to become
  established.

Toolbox

Overview and questions

Tools	Page	Questions
Success factor Strategy		
1. Strategy Suite	94	What steps can be used to develop a project's strategy?
2. Societal Patterns and Trends	99	What social developments and trends influence the project?
3. Scenarios	102	How does the area of social concern develop over time?
4. Key Challenges: SWOT	106	What strengths, weaknesses, opportunities and risks shape the area of social concern in which the project aims to achieve results? What capacities are available at the different levels of capacity development?
5. Devising Options	110	How can you develop alternative options for the project strategy?
6. Selecting an Option	113	How do you choose the right project strategy?
7. Results model	116	How are the intended results of a project linked? What are the underlying hypotheses for these links?
8. Capacity Development Strategy	122	What capacity development strategy looks promising given the project's intended objectives and results?
Success factor Cooperation		
9. Map of Actors	129	What individuals and organisations are (or should be) involved in the project?
10. Actor Profiling (4 A's matrix)	134	How do the actors' profiles look in detail? What does this mean for cooperation within the project?
11. Interests of Key Actors	137	What interests link the key actors with the project? Where does conflict arise? How can you deal with this? How do you deal constructively with conflictual relationships or conflicting interests?
12. Structural Characteristics of Cooperation	141	What characteristics can you use to measure the quality of cooperation? What patterns of cooperation are suitable? What roles do the actors adopt in the project?
13. Views of Actors (PIANO Matrix)	149	What do the key actors have in common with the project? How can the project link into the actors' strategies for action?
14. Networks: Strengthening Relationship Potentials	153	When does it make sense to set up a network? What key questions arise as part of network management?

Tools	Page	Questions
15. Trust-Building	158	How can you consolidate cooperation relationships? What tensions and conflicts are evident? What tools, conflict management systems or conflict resolution approaches can you use to address this?
16. Backstage and Learning Behaviour	165	What implicit rules, problem-solving approaches and learning patterns shape the cooperation system?
17. Needs Analysis	171	In which project areas is complementary cooperation needed? With which actors could you potentially establish a cooperation partnership?
18. Comparative Advantages	174	What comparative advantages make the project an attractive partner in complementary cooperation?
19. Shaping Negotiation Processes	177	How can you structure negotiations in which different interests, ways of working and expectations meet and perhaps clash?
Success factor Steering structure		
20. Steering Structure	182	How does the project make decisions? How should the different actors be involved in steering? How will a suitable steering structure for the project look?
21. Qualities of a Steering Structure	187	What requirements should the steering structure meet?
22. Results-Based Monitoring System	190	How do you set up a results-based monitoring system? What do you need to take into account during set-up and operation?
23. Architecture of Intervention	194	How do you design interventions over time?
24. Plan of Operations	200	How do you channel the strategic orientation into an implementation plan?
Success factor Processes		
25. Process Map	205	What processes exist in the area of social concern or in the project?
26. Process Hierarchy	210	How do you break processes down to an operational level?
27. Process Design	213	How do you design and describe processes in detail?
28. Process Optimisation	216	How can you recognise and optimise critical processes?
29. Interface Management	221	How do you design interfaces?
Success factor Learning & innovation		
30. Scaling-Up	225	How can you shape the process to scale up innovative approaches?
31. Learning Capacities in Cooperation Systems	229	How do you review and adjust learning needs and learning results within a project?

Tools	Page	Questions
32. Innovative Capacity of Cooperation Systems	232	How can you build a project's innovative capacities?
33. Knowledge Management in Projects	236	What knowledge products should you document? How can you do this successfully?
34. Debriefing	241	How do you evaluate and safeguard the lessons learned in projects?
35. Learning Networks for Multipliers and Trainers	245	How can you structure institutional exchange among multipliers and trainers?
36. Communities of Practice	249	How do cooperation systems and organisations learn from experience and existing know-how?
37. Organisational Diagnosis	253	What is an organisation's performance capacity? What learning needs does it have?
38. Quality Management in Organisations	258	How can you systematically improve the quality of output delivery in organisations?
39. Quality Assurance in Competence Building	263	What aspects do you need to take into account when designing effective training strategies?
40. Intervision	267	How can implicit knowledge that has been acquired by specific individuals be made available and exchanged by actors?
41. Developing Learning Objectives	270	What learning objectives does the project aim to achieve? What tools will help you draw up these objectives?
42. Reviewing a Project Learning Strategy	274	At which levels do you need to develop learning capacities to a greater degree? In which areas?

Success factor Strategy

Tool 01

Strategy Suite

Notes on use

Purpose	This suite of tools will help you devise strategies for capacity development or for projects that aim to support change in social sub-systems. It offers a series of tools ranging from analysis to the development, evaluation, and decision on a strategic option. It delivers a properly formulated strategy.
When to use it	In situations in which the actors involved are reviewing or clarifying a project's strategic orientation; at the start of a project and possibly during a project phase when monitoring information calls the strategy into question, when objectives change or when the context of the project changes.
Setting	Different sized groups; given the structure of the tools, we recommend that a defined group of individuals be able to work through the sequence from start to finish in one workshop or in two consecutive workshops.
Facilities and materials	Pinboards, flip charts, workshop materials (markers, cards, pins etc.); photographic documentation of all outcomes.
Notes	It is crucial to start with a clearly defined strategic issue. The tools are sequenced in a way that guides you through issues, getting you to compare different hypotheses and ideas. The strategy suite combines tools from all of the success factors. By delivering a properly formulated capacity development strategy, it provides you with a point of entry into the other success factors, which are then structured accordingly for project management based on the option selected. We recommend that you use an external moderator or consultant for this process. He/she takes on responsibility for processes and ensures that the targeted outcomes are achieved.

Description

Developing a strategy means working through a process. To help you structure this process, Capacity WORKS offers a recommended sequence of steps and corresponding tools in the strategy suite. In each step, you can choose from and in some cases combine a variety of tools, depending on your specific needs.

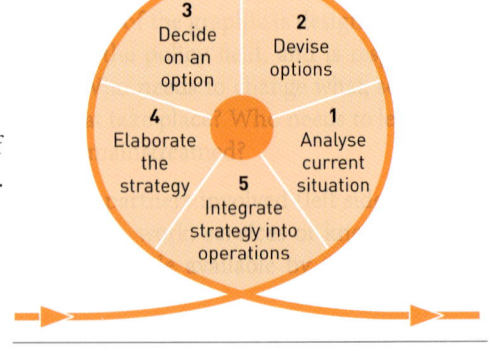

Figure 17: The strategy loop

How to develop a strategy is best explained by outlining a process that breaks the strategy loop down into a linear sequence of steps as follows:

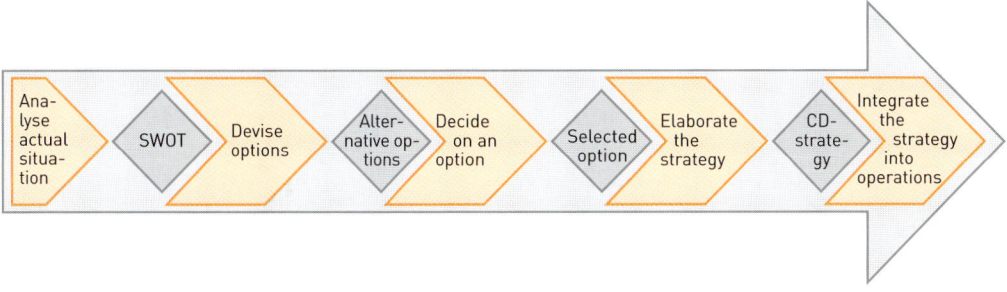

Figure 18: Strategy development as a linear process

Organisation of the process

As we can see in Figure 18 above, the strategy development process comprises five steps, each of which has its own tools. Each process step produces an outcome using one or more of these tools. Experience shows that it usually takes several workshops (which need not take place on consecutive days) to work through the process. You can schedule a break or time for reflection after each process step and each outcome achieved. However, it is also important that you maintain a sufficient flow throughout strategy development to sustain the impetus for innovation within the group and leverage the momentum inherent in each step. Remember to carefully document each step and each outcome so that they can be explained to anyone not present at the workshops. In this way, you will be able to ensure that the rationale for the strategy is recorded transparently and can be discussed at a later stage in the project if necessary.

When developing a strategy, it is important that you:

- define the actors to be involved in strategy development in the steering structure and in the responsible bodies. Key actors who make relevant decisions and contribute resources are particularly important in this context, as is involving a variety of actors to ensure that you incorporate the widest possible range of perspectives;

- consider hiring a consultant to moderate proceedings so that all actors are free to contribute and remain focussed on outcomes throughout the process.

- think about how those actors who will be affected but were not involved in the development of the strategy can be informed or included later on;

- stipulate how outcomes can be clearly documented and perhaps used to present the project for PR purposes, for instance.

Key questions for preparing strategy development:

- Why is the strategy being developed?

- What key issues will the strategy focus on? Do these issues affect the entire project or just part of it? What actors does it affect?

- Who should be involved in strategy development?

- Do some strategic elements already exist or have they already been described in a previous phase of the project? Do you wish to re-use them?

- In which regional and political framework does the development of a strategy for the project fall? What documents that determine the boundaries or general conditions should be incorporated?
  (possibly relevant for German international cooperation projects: priority area strategy papers published by the German Federal Ministry for Economic Cooperation and Development (BMZ), BMZ country strategies, national sector strategies, supraregional strategies, strategy papers published by other donors, NGOs etc.)

- Who will moderate the whole process, who will provide consultancy and take responsibility for the process and its preparation?

- When will you look at the strategy and use it? When will you update it?

We recommend that you always develop a strategy at the start of a project. Throughout the calendar year, it makes sense to synchronise drafting with the reporting schedule and with the monitoring and budget processes of the main parties involved. You must draft a strategy in order to provide a rationale for planning activities and allocating resources in milestone and operations planning. When scheduling the processes involved, remember that your strategy is closely interlinked with your resources and operational planning.

Previous experience shows that you need to review the strategic orientation once a year. A change in government, an economic crisis or conflict could spark a review, as could upcoming evaluations or appraisals, which often tend to be prompted by the internal processes of one of the actors involved.

The reason for drafting a strategy often determines its scope: Is it needed for the entire project or only for a specific area? Is it being drafted at the start of the project for all aspects of the project? Or is it being drafted at a later stage to review specific aspects of the project?

Once you have compiled the key issues for strategy development together with the participants, we recommend that you write them down e.g. on a flip chart so that they remain visible throughout the entire process.

Having established the focus of strategy development – i.e. what key issues will you address –, defined the group of actors to be involved and clarified the organisation of the process as well as the consultant to be involved (where necessary), you can move on to the following steps.

How to proceed

The following table provides an overview of the five steps involved in strategy development and the tools that may be used in each step. Some of the tools are described in other success factors, in which case we only refer to them briefly here. Others form part and parcel of the success factor described here and are explained below.

Process steps	Toolkit	Tool number in the SF	Outcome produced by the process step
Step 1 Analyse current situation	**Societal patterns and trends** Scenarios **The process map** Actor profiling (4 A's matrix) **Map of actors** Comparative advantages Needs analysis **Key challenges: SWOT analysis**	02 03 25 10 09 18 17 04	▪ SWOT
Step 2 Devise options	**Devising options**	05	▪ Alternative options
Step 3 Decide on an option	**Selecting an option**	06	▪ Selected option
Step 4 Elaborate the strategy	Results model **Capacity development strategy**	07 08	▪ Results model ▪ Capacity development strategy
Step 5 Integrate strategy into operations	The outcomes produced by strategy development are channelled into the other success factors		

Figure 19: Strategy development toolkit

In order to effectively apply the strategy loop, we recommend that you always use certain tools, which are highlighted in bold above. You can apply any of the other tools in the relevant steps selectively as necessary, depending on the specific issues to be addressed.

The steps are logically sequenced and will help you switch between broadening and narrowing your focus as you follow the loop. The information gathered during analysis will initially make things more complex, before you proceed to condense it into key challenges. By devising several alternative and ambitious options in a playful manner, you will gain a broader view of the potential scope for action. Selecting an option based on a transparent rating awarded by all the participants reduces the complexity by focusing on an action-oriented decision. Based on this decision, the selected option is then further elaborated by developing interventions, outputs or the results model.

Use this toolkit to compile a sequence of step-by-step tools, and apply them together with the actors involved as you work through the loop in a workshop setting.

Tool 02
Societal Patterns and Trends

Notes on use

Purpose	This tool allows you to conduct a brief and pragmatic analysis of the overall societal context, which is relevant to determining the focus or key issues for strategy development. It will clearly identify societal patterns and trends and interpret them as opportunities and threats for achieving the project's objectives.
When to use it	In situations in which you need to examine and review in greater detail the political, economic and social context in which the project is embedded.
Setting	Workshop (the smaller the group the better; up to twelve participants) with actors with local know-how and – most importantly – a good knowledge of the societal context.
Facilities and materials	Pinboards, flip charts, moderating materials (markers, cards, pins, etc.)
Notes	This tool adopts a pragmatic approach. A small group can work through the process in two hours. It is also a suitable tool for individuals who wish to take a structured approach to analysing the actual situation. If, however, you wish to conduct a more detailed analysis of patterns and trends, other methodological approaches such as GIZ capacity assessment are more suitable for examining the actual situation in greater depth.

Description

Each project in a cooperation system is embedded in a general societal context. The key issues or focus orient strategy development toward the most relevant aspects of the sector or social sub-system. However, it is important to routinely review how a project is embedded not only within the sector, but also within the overall societal context in order to ensure that a project's strategic orientation is appropriate and that it is on the right path for change. We understand the project context as a broad array of forces that interact with each other and affect the project. To ensure that a project is appropriately embedded in the context the relevant forces must be identified as trends and/or patterns in sub-systems of society, such as the economy or education, and the project's strategy must be explicitly oriented toward these.

Rather than focussing on the immediate project environment – which is usually examined in actor and process analyses – a contextual analysis looks at the broader environment, which spans the general societal, national or supranational (for supraregional or global projects) and political context. Although this broader environment influences the project, the project may find it difficult to exert influence on it.

The overall context of a project can be viewed from four different structural perspectives:

(1) The social perspective
This perspective looks at social sub-systems – such as culture and religion, mass media, education, family, and civil society. Factors such as demographics, cultural attitudes, values, lifestyle, life expectancy, illiteracy, etc. play a key role in this context.

(2) The technological perspective
This comprises the sub-systems of science, technology, and research.

(3) The economic perspective
Here, patterns and trends are examined against the backdrop of the relevant economic model (e. g. a planned economy, a social market economy, neoliberalism); the steering of economic development by public and private actors; the financial, managerial and operational resources of the private sector; the ability of the private sector to organise itself (associations, chambers etc.); the importance of the informal sector; the relationship between actors from the public and private sectors, or access to financing and credit.

(4) The state perspective
This perspective looks at trends and patterns in the political system and in the political decision-making process. These include: the structure of the administration and the understanding of governance; the legal framework, the separation of powers, the enforcement of legal standards, scope for civic participation in the political system; national and transborder conflicts; fragility; key development strategies by the state. Issues such as environmental protection, climate change adaptation, sustainability, anti-corruption, and health care and education as public services could also fall under this category.

How to proceed

Step 1: Identify patterns and trends

In a brainstorming session, start by listing trends and patterns in the identified fields. Assign each idea to a relevant field.

If the brainstorming session is guided and observations are being corroborated, we recommend that you work through the following questions:

Question 1: What current trends in the societal context could be relevant for the project?

Question 2: What patterns are prevalent in the societal context?

Question 3: What events corroborate the relevance of the trends and patterns? How volatile are they?

At the end of this step, you will have documented the identified trends and patterns.

Step 2: Describe the effects that patterns and trends have on the project

In this step, you shed light on the identified patterns and trends by asking questions that help pinpoint and describe the relevant factors that influence the project.

- How do the patterns and trends that could be relevant affect the project?
- What negative or positive effects are to be expected?
- What patterns and trends have a neutral effect or are irrelevant, irrespective of the strategic path the project adopts?

Step 3: Understand interdependencies

The following question will help you identify links between the patterns and trends.

- How are the patterns and trends linked?
- Do any of the patterns and trends conflict with each other?
- Do any of them reinforce each other?

Step 4: Focus on relevant effects and how the issues will develop

By now, you may have identified many different patterns and trends. If so, you can whittle them down to a few that are particularly relevant for the issue addressed by the project.

Optional:
To be able to forecast the trajectory of the relevant issues or patterns you usually need to analyse the driving forces that underlie them. Which forces reinforce the patterns and issues? Which ones limit or hamper them? It is often helpful if you run through different scenarios of how issues could develop.

Step 5: Draw conclusions

The patterns and trends you have identified in the previous steps provide a framework for the project. They describe key challenges that you need to take into account when defining the project's strategic orientation. You can use a SWOT analysis to depict them as strengths, weaknesses, opportunities or threats. (For more information, see the tool 'Key challenges: SWOT').

You could formulate alternative possible responses to strengths, weaknesses, opportunities and threats, which could include the following:

- Avoid cooperating with XY
- Avoid the issue XY
- Make sure that you look at the process XY in the sector
- Observe trends closely
- Take into account decision-making patterns in the institutions XY

Tool 03

Scenarios

Notes on use

Purpose	Use this tool to assess – through an exchange of different perspectives and experiences – relevant factors and their effects on future developments. This assessment will provide you with a better basis for decision-making.
When to use it	This tool allows you to devise different scenarios for how issues might develop. It may be used as a follow-up to the tool 'Societal patterns and trends', and is based on a snapshot of the area of social concern at a particular moment in time. Against a backdrop of potential 'best' and 'worst-case' scenarios, you assess the probability that certain developments will occur. Scenarios can help you formulate an overall strategy as well as sub-strategies.
Setting	Workshop with key actors
Facilities and materials	Pinboards, workshop materials (markers, cards etc.); pre-prepared scenario cone, handouts of the relevant documents.
Notes	Before the workshop, it is a good idea if you conduct research on trends for relevant factors. Remember that even highly sophisticated scenario analyses can only provide an indication of the likelihood that the events will actually occur. The main aim here is to shed light on and open up scope for discussing the perspectives of different participants. To ensure a balanced approach it is helpful if you include both highly pessimistic and highly optimistic individuals in the analysis.

Description

Scenarios will help you describe and compare various paths toward future development. Images and models of possible future developments are useful for exploring various options for action. Unlike forecasts, scenarios do not attempt to predict the future unequivocally, but seek to identify possible future events and developments. These are approximations, based on the existing knowledge and experience of the participants who devise the scenarios. In other words, scenarios create a pragmatic link between the uncertainty of the future and the need to take decisions today.

Thinking in scenarios is a conscious attempt to face the basic unpredictability of the future, to identify trends and developments, and to reflect on the consequences they have for projects. Scenarios are based on the simple fact that the future is uncertain, and involves risks and potentials about which we are unable to say anything definitive.

Scenarios change the participants' perspective:

- They enable participants to approach the area of social concern from a more comprehensive and differentiated perspective.

- They provide concrete points of reference and stimulate discussion.

- They support a systemic perspective on the area of social concern.

- They broaden each participant's own perspective by incorporating a wide range of different perspectives.

- They structure and promote an exchange of perspectives and experiences among the participants.

How to proceed

Step 1: Define the area to be analysed and the time frame

Start by defining the area of social concern to be examined, along with the time frame.

The starting point of a scenario is always the present. The time frame is defined as the projected interval between the present and a specific point in the future, for instance in four or ten years' time.

Step 2: Identify factors

In this step you identify variable factors that affect future developments in the area of social concern.

Collect concise statements concerning the following points:

- socio-economic and political and institutional trends in the area of social concern;

- important action strategies of different actors;

- possible events that may significantly affect future developments.

You now systematically assess the statements you have collected from the following perspectives to identify relevant factors and their effects on future developments (cf. The tool 'Societal patterns and trends' for more information):

- the social perspective;

- the technological perspective;

- the economic perspective;

- the state perspective.

Draw up a list of the relevant factors you have identified in the area of social concern.

Step 3: Evaluate the factors

Rate the identified factors in terms of their importance and the probability that they will occur. Assign each of the individual factors to a field in the following matrix:

Importance	High	Volatile trends and key factors (including negative factors)	Major known factors that must be taken into account
	Low	Volatile trends that have little effect right now	Factors that today are largely known but have no effect
		Low	High
		Probability	

Working aid 1: Matrix of factors

The factors can be rated in a number of different ways. Either they can be rated by each participant individually and the assessments aggregated, or the workshop participants can agree to first discuss the factors and then produce a joint rating.

Step 4: Define the main factors

You can now pinpoint what main factors will influence future developments. In this context, focus on the factors that were deemed important in the previous step. If you identify more than six factors, whittle these down to between four and six main factors.

Step 5: Formulate contrasting scenarios

In this step, participants formulate two coherent, plausible visions of the future in the form of two contrasting scenarios – a best-case scenario (scenario A) and a worst-case scenario (scenario B) – for the main factors that will influence future developments. The scenarios should be documented in writing, and where possible illustrated using images and/or given a succinct title. The two scenarios form either side of a cone, the pointed end of which is located in the present. In other words, the further we look into the future, the greater the degree of uncertainty (i.e. the wider the cone).

Scenarios

- … are visions of alternative, consistent, future situations. Each scenario presents a vision of a possible future that is plausible (that can happen), coherent (that is logical) and credible (that can be explained).

- … are accounts of possible future courses of events and situations based on currently identifiable trends and ideas about the future.

- ... sharpen our awareness of potentials and risks. They expose and bring into focus our assumptions about future developments, and about the driving forces behind them.

- ... illustrate complex projections, and make them easy to grasp.

- ... generate a creative climate, and enable us to think in terms of alternative outcomes and scope for action.

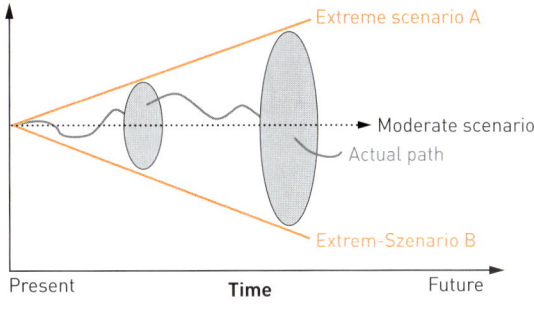

Figure 20: Contrasting scenarios

In addition to formulating 'best' and 'worst-case' scenarios, you may choose to draft a third 'probable' scenario.

Step 6: Draw conclusions

(1) Conclusions as regards key challenges in the area of social concern:
Devising scenarios is based on the assumption that formulating best and worst-case scenarios will help you get a good feel for the conditions that will shape the area of social concern over time. So, in addition to analysing societal patterns and trends, scenarios will help you get a handle on strengths, weaknesses, opportunities and threats in the relevant area. You can then analyse these using the 'Key challenges: SWOT' tool.

(2) Basis for identifying and describing options:
You can also draft scenarios in the run-up to devising options (cf. The 'Devising options' tool). The actors who worked together to draft scenarios will have created a sound basis for devising, discussing and describing options.

(3) Criterion when rating options: resilience
Rating existing options (cf. the tool 'Selecting an option') constitutes an important step in strategy development. You can use best and worst-case scenarios as a criterion for resilience during the benefits analysis. For example, you could assess how successful the identified options would fare against the backdrop of the best and worst-case scenarios. The average value would indicate the resilience of the various options, allowing you to deduce how robust a particular option would be in a given context.

Tool 04
Key Challenges: SWOT

Notes on use

Purpose	During strategy development, this tool will help you to structure the relevant challenges in the area of social concern in which the project is to support change. It: ■ analyses the strengths and momentum for change that can be leveraged by the project; ■ analyses the weaknesses that the project is to address; ■ identifies opportunities in the project setting that can be used to shape the change process; ■ identifies the threats that jeopardise the need for change
When to use it	To structure the findings of the analysis of the actual situation. A SWOT analysis will help you to summarise and identify the most salient points identified during the analysis. It will also prove useful in assessing the strategic options you develop in the next stage.
Setting	All participants involved in strategy development. (Can be used in a range of different settings).
Facilities and materials	Pinboards, workshop materials (markers, cards etc.); handouts of the relevant documents.
Notes	The outcomes produced by all activities carried out in the strategy development process to date (e.g. process map, analysis of actors, societal patterns and trends, hypotheses from the analysis of the actual situation) should be available and pre-processed. In the discussion itself, it may be useful if you are able to flesh out a previously drafted proposal of the strengths, weaknesses, opportunities and threats (for example, if all participants are not familiar with the findings of the analysis of the actual situation). The key challenges should be structured immediately before you develop the strategic options. These two steps are closely interrelated, so it is important that you perform them in quick succession.

Description

The key challenges to be faced in the area of social concern provide you with some indication of the capacities that will be required in the permanent cooperation system. 'Capacity' is the ability of people, organisations and societies to manage their own sustainable development processes. This includes recognising development problems, designing strategies to solve them, and then successfully implementing these. This ability is often also referred to as the ability for proactive management, which refers to people's capability to effectively combine and coordinate political will, interests, knowledge, values and financial resources in order to achieve their own change objectives and needs.

The quality of the information you elaborated during the analysis of the actual situation will ultimately determine the quality of the capacity analysis. This information paves the way for subsequent stages where it is analysed and structured against the backdrop of societal patterns and trends.

Your aim here is to break down the information you collected during the analysis of the actual situation and categorise it into strengths, weaknesses, opportunities and threats. On this basis, you then draw conclusions for the (future) project. This includes an assessment of the capacities in the permanent cooperation system, which will allow you to derive sound strategic options for a temporary cooperation system in the stages that follow and to develop a capacity development strategy for this system.

How to proceed

You will need the findings and hypotheses established during the analysis of the actual situation to compile a SWOT analysis. This information may need to be pre-processed and structured to some degree, depending on the type and number of participants involved. In the case described below, we recommend that steps 1 and 2 be prepared in advance.

Step 1: Answer the question 'capacities for what'

To assess capacities you will need to identify a contrasting situation that you can use as a yardstick to pinpoint where you need to be. Answer the question 'What does the society or area of social concern need the increased capacities for?' Identify this situation from the perspective of the permanent cooperation system. Based on this perspective, how would you describe the target situation in a nutshell? In other words, from today's point of view, where do you as actors within the area of social concern need to get to? And what capacities do you need to do this?

For example, if you want to conduct an educational campaign to prevent the spread of certain diseases, you will need different capacities than if you need to improve the performance of public hospitals. The capacity assessment for each scenario will vary accordingly. So too will the capacity development strategies you devise.

Therefore, if you want to achieve targeted outcomes using a SWOT analysis as proposed here, you will need to establish a clear focus. To do this, ask yourself the following questions. They will help you home in on your goals and steer any subsequent discussions toward the momentum for change inherent in the area of social concern:

- Based on the findings of the analysis of the actual situation, what degree of consensus has been reached at the societal/political level regarding the future structure of the area of social concern?

- Are any initiatives for change supported by a critical mass of actors?

Break down your hypotheses into the different capacity development levels (working aid 2). Use them to guide further discussions. Assuming that the process map provides a clear overview of the processes involved in the area of social concern or sector in its current state, you could visualise the answers to these two guiding questions in the process map (e.g. by modifying selected steering or support processes, establishing a new learning process). The resulting process map will depict the vision for the area of social concern.

General description of the permanent cooperation system based on the current perspective: For what target situation do you need to build capacities? *(visualise so that all workshop participants can see)*	What **consensus** on the future structure of the area of social concern is evident and seems realistic at the **societal/political** level? (Draft a rough outline of the trends observed, based on the findings of the analysis of the actual situation).	What **initiatives for change** that are supported by a critical mass of actors are already evident that a project could promote? (Brief description of issue, cooperation partners, financial framework, assessment of feasibility)
Level of society – enabling frameworks		
Level of society – cooperation systems		
Level of organisations		
Level of individuals		

Working aid 2: Capacities for what?

Step 2: Analyse the strengths, weaknesses, opportunities and threats presented by the findings of the analysis of the actual situation

In this step, you consolidate and summarise the findings of the previously conducted analysis of the actual situation. Break down the conclusions you drew there into the different capacity development levels and assign them to the categories Strengths, Weaknesses, Opportunities and Threats. It will help if you visualise them using the tables shown in working aid 3 below.

Strengths	
What strengths are evident for the area of social concern?	
Society: Enabling frameworks	
Society: Cooperation systems	
Organisation	
People	

Opportunities	
What opportunities are evident for the area of social concern?	
Society: Enabling frameworks	
Society: Cooperation systems	
Organisation	
People	

Weaknesses		Threats	
What weaknesses are evident for the area of social concern?		**What threats are evident for the area of social concern?**	
Society: Enabling frameworks		Society: Enabling frameworks	
Society: Cooperation systems		Society: Cooperation systems	
Organisation		Organisation	
People		People	

Working aid 3: SWOT analysis

Step 3: Discuss the findings

The findings of the SWOT analysis should help you answer the following questions: What key challenges will you need to address when developing and assessing strategic options in the subsequent stages? Within the area of social concern, what capacities already exist that support the social and political consensus on the future outline of the area of social concern and the existing initiatives for change? In other words, always discuss the strengths, weaknesses, opportunities and threats against the backdrop of the capacities identified in step 1. N.B.: If you have drafted a SWOT analysis prior to the discussion in a workshop with the relevant actors, do not omit step 1 but first describe what the additional capacities are specifically required for, reach a consensus on them and then proceed to step 3 and start the discussion on the SWOT analysis.

Breaking hypotheses down into the different capacity development levels will help you keep a close eye on all dimensions of the area of social concern. This breakdown should be used actively by the facilitator to guide the parties involved through the discussion by, asking the following questions, for example:

- At the societal level, which strengths or opportunities are conducive to achieving the targeted change?
- At the societal level, which weaknesses or threats could hamper achievement of the targeted change?
- At the organisational level, which strengths or opportunities are conducive to achieving the targeted change?
- And so on and so forth for all levels

N.B.: If a SWOT has been pre-processed by a preparatory working group we recommend the following procedure: Start with a presentation of the SWOT analysis for the whole group. Where the group's size and the background knowledge of all participants permit, we recommend breaking the group down into smaller working groups to discuss the pre-prepared SWOT analysis and supplement or amend it accordingly. This should be followed by a discussion within the entire group. The discussion should aim to reconcile any different views on the strengths, weaknesses, opportunities and threats in the area of social concern, at the societal (including enabling frameworks and cooperation systems), organisational and individual level. Through the discussion participants will arrive at a joint version of the SWOT and a shared picture of the key challenges.

Tool 05
Devising Options

Notes on use

Purpose	This tool will help you devise possible strategic options for the (future) project, which will describe how to achieve your targeted change objective in the relevant area of social concern.
When to use it	To spark discussion on the different ways in which you can achieve the change objective and, together with other actors, come up with well-thought-out ideas rather than hasty 'blueprints'. It will help you identify alternative solutions and reach a conscious decision on a single strategic focus.
Setting	Workshop with key actors
Facilities and materials	Pinboards, workshop materials (markers, cards, etc.) Relevant documents: The outcomes produced to date should be available in a visual format (e. g. SWOT analysis, process map, map of actors and societal patterns and trends).
Notes	You should discuss strategic options directly after you discuss the key challenges (which are mapped in a SWOT analysis). In this way, you will have a clear picture in your mind's eye of the current status of the area of social concern. It is important that you provide a creative and open atmosphere that is conducive to devising a wide variety of options. During this stage, it is important that you generate and document as many ideas as possible.

Description

Strategy development involves identifying strategic options that will enable you to choose the most promising path for change in order to achieve the project's targeted results. Identifying a number of options allows you to think 'outside the box'. You can work through the different alternatives against the backdrop of the relevant area of social concern and free yourself from prescribed blueprints. Use this tool to identify all options that appear viable, based on the information gathered during the analysis of the actual situation.

This approach will open up additional scope for finding creative solutions and allow you to avoid: (1) using 'one-size-fits-all' strategies that may be state-of-the-art but are unrealistic in the given context; (2) having to work with a number of different options simultaneously, which ultimately prevents you being able to focus resources and establish a clear project profile ('all show no substance').

How to proceed

The following ideas may help you formulate strategic options and encourage those involved to be creative:

- Strategic options describe different ways of achieving a particular objective.

- Only by subjecting fixed articles of faith and assumptions about how the area of social concern works to a critical review, can you draft creative and innovative options and avoid looking to the future through the rear-view mirror.

- Try to quell any urge you may have to rip a new idea to shreds. When devising different options give everything a chance, however unrealistic or unfeasible an idea may seem at first glance.

- Devising strategic options can be deemed a success if you are able to come up with suggestions that no one would ever have dreamed of at the outset.

Step 1: Form creative groups

If there are enough participants, create several small groups of between two to four people per group. Those involved will come from a variety of backgrounds and will have participated in the preceding discussions.

- Have each group draft up to three options.

- Individual warm-up: Get each member of the group to brainstorm initial ideas as keywords for a few minutes.

Step 2: 'Walk & talk' in groups

During a walk (preferably out of doors, in an environment other than the workshop room such as a park or garden), the participants within each group share their first keywords. By channelling their powers of association, the groups will develop initial ideas for strategic options. This step will take participants out of the familiar workshop routine and create greater scope for creativity.

Step 3: Visualise ideas

After the walk, each group uses a flip chart to document their ideas. For each idea, it outlines a heading, characteristics and a symbol or image that represents the idea. Keywords – to describe products or the general thrust of a strategy, for example – are an important instrument for describing options.

Step 4: Present findings in the 'gallery'

All groups meet to briefly present their findings. Ideas are not yet discussed, but questions can be asked if something is unclear.

Step 5: Summarise findings

In this step, you pool similar options. Consolidate the findings in a way that clearly describes the remaining options. Whittle down the strategic options identified to between three and seven possibilities. Ensure that people can relate to the options identified. On the one hand, they should have an appropriate level of complexity that is in keeping, for example, with the complexity of the change objective. On the other, they shouldn't be too fragmented or 'high-brow'. The different options should be on the same level of abstraction so that they are comparable.

Step 6: Describe the strategic options

In this final step, you describe the details of the strategic options devised.

What would it mean if this option were pursued? What work packages would be developed as a result? Who are the most important actors? What processes (output/cooperation/learning/support and/or steering processes) can be used as starting points?

N. B.: You must draft a description of potential strategic options before moving on to assess them and make a final selection. Otherwise each participant will associate different features with them, and there will be a lack of any shared and more precise understanding.

The following table (working aid 4) will help you depict the key characteristics of each option in a uniform manner:

Strategic options	Description			
	Symbol/image	Work packages	Key actors	Processes (output/cooperation/ learning/support and/or steer- ing processes) that provide a good starting point
Heading for strategic option 1				
Heading for strategic option 2				
Heading for strategic option 3				
...				
...				

Working aid 4: Description of the strategic options

Tool 06
Selecting an Option

Notes on use

Purpose	This tool will help you conduct a structured discussion to assess strategic options and to come to a well-informed decision.
When to use it	Once you have identified diverse options.
Setting	Workshop with key actors.
Facilities and materials	Pinboards, workshop materials (markers, cards, etc.) Visualised description of the strategic options. Visualised results from the analysis of the actual situation (e.g. map of actors, process map, and societal patterns and trends).
Notes	Ownership can only develop and decisions can only be made jointly if the relevant supporting actors are involved and if the process is well designed. The development and assessment of different strategic options and a decision on one of them involves a negotiation process. Discussions may often be complex, detailed and sometimes arduous. It is important to spot tendencies to avoid honest discussion and the struggle to reach a joint decision. It is often helpful if you draft observations on possible criteria for discussing and assessing strategic options before the workshop.

Description

Once different strategic options have been identified, the participants set about making a joint decision on which strategy to pursue. Ask the following questions:

- What criteria will be used to assess the different strategic options?
- What are the advantages and disadvantages of the different options?
- What results and risks are anticipated for the individual options?
- Which option seems the most promising?

How to proceed

Step 1: Agree on the assessment criteria

To ensure that you choose the most realistic option, make sure that all of the information gathered up to this point from the analysis of the actual situation is taken into account. If this information is available in a visual format, make sure that this is available too or can be viewed by all of the participants, who should have a good idea of the key characteristics of the strategic options up for discussion.

The assessment criteria will vary, depending on the context, and should be agreed between the different participants. They could include:

- leverage within the area of social concern;
- willingness of key actors to change;
- the feasibility of the option against the backdrop of the societal patterns and trends identified;
- the feasibility of the option against the backdrop of the existing capacities at the three levels of capacity development;
- compatibility of the options with the project's strategy;
- the sustainability of results in the permanent cooperation system;
- scalability;
- risk probability;
- resilience (how robust is the strategy given the ambient factors?);
- the funding required;
- synergies with other actors;
- degree of use of available expertise;
- compatibility with demands of a particular commissioning party.

This list of suggested criteria will provide you with a general basis for tweaking a specific scenario, and enable you to develop ideas you may have about possible criteria. Collect other proposed criteria and discuss and agree on them. We recommend stipulating no more than between five and seven criteria. You could also weight the criteria, if desired.

Where possible, assign benchmarks for the criteria in the following working aid and display it on a pinboard where it is visible to all:

	Assessment criterion A	Assessment criterion B	Assessment criterion C	Assessment criterion D	Assessment criterion E	Etc. ...
Strategic option 1						
Strategic option 2						
Strategic option 3						
Etc. ...						

Working aid 5: Criteria for selecting an option

Step 2: Assess the strategic options

Use the above matrix to assess the different options identified. Rate the options using a 'traffic light' system or on a scale (e.g. from 0 to 10 or from 1 to 5). Use a rating system that works for the participants.

Discuss all of the criteria option-by-option and document the ratings in a way that is visible for all participants. Using this approach will help you to focus on the strategic options as a group and will help you to clarify the priorities, common ground and differences that are important for making a joint decision.

Step 3: Weigh up the effectiveness and the risks

You can evaluate the different options in an effectiveness/risks matrix in addition to using the detailed assessment matrix shown above. To do this, the options are mapped on a coordinates system with two axes; one depicting the possible effectiveness, the other the potential risks. The possible combinations of both categories produce four quadrants:

Ideal: Effective results with low risk: ideal scenario

Risky: Effective results with high risk: risky scenario that requires precautionary risk-management measures to be taken and cut-off points to be defined

Irrelevant: Less effective, low-risk results: these options represent less-relevant alternatives that could still be implemented, however

No go: Less effective, high-risk results: these options are to be avoided

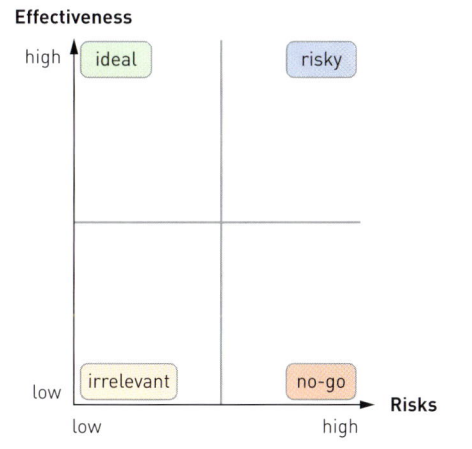

Working aid 6: Effectiveness/risks matrix

Mapping the strategic options to the quadrants involves a process of negotiation that should be based on the findings of the assessment matrix. In this step, the participants should agree on where to position each option.

Step 4: Decide on a strategic option

Aggregate the ratings awarded in the previous steps and select the option that performed best for all of the criteria rated.

As the discussion unfolds it may prove useful to combine specific elements of different options in one in order to better address the risks identified, for example.

N. B.: In cases where relevant actors are not actively involved in the process, we recommend that you identify a range of acceptable options that will provide you with a better basis for any further negotiation and decision-making required.

Tool 07
Results Model

Notes on use

Purpose	This tool depicts the selected strategic options at a glance. It will help you to clearly document a joint understanding of the path to change for all actors.
When to use it	To work through the strategy development process and to negotiate realistic objectives within the context of the cooperation system. While working on the results model activities are agreed that will help achieve the intended changes. So too are the relevant contributions the individual actors of the cooperation system will make to generate the activities.
Setting	The specific setting will depend on how the strategy development process is designed.
Facilities and materials	Pinboards, workshop materials (markers, cards, etc.) Description and/or visualisation of the strategic option selected. Visualised outcomes of the analysis of the actual situation.
Notes	Before you use this tool, you need to decide on a strategic option. In other words, it must be clear what approaches the project will pursue within the area of social concern.

Description

Cooperation systems channel momentum from the jointly agreed objectives and from the diversity of the actors involved. Objectives are embedded in the consensus reached between the actors regarding the area of social concern in which the intended results are to be achieved through joint activities.

You can use a results model to map the agreed objectives. The results model presents a progressive sequence of causally interdependent positive changes. It depicts a change process that will be supported by jointly agreed activities. As a *model*, it is a simplified representation of reality. It does not claim to represent the full complexity of the actual situation.

Using a results model fulfils several functions. It …

- assures quality during strategic planning. The results model summarises the strategic orientation and the conceptual design of a project.

- clarifies the areas that activities will address.

- provides a guideline for joint steering. The actors base their approach on the underlying results model, which they use as a basis for implementation.

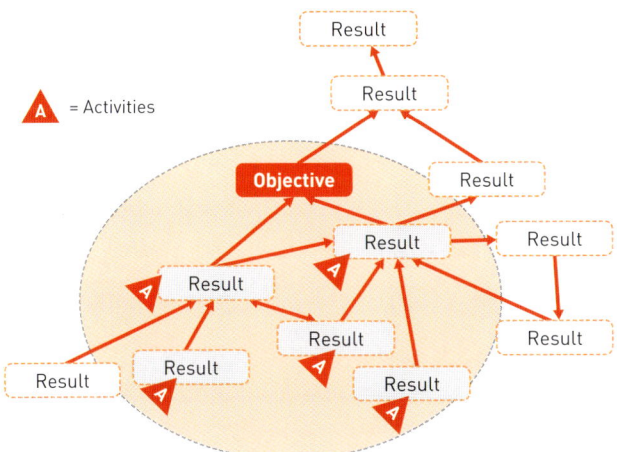

The results model maps the entire change process in a sector and shows the entry points of an intervention.

Figure 21: Results model

- provides a basis for results-based monitoring.

- helps when preparing reports on results for the relevant actors.

- is the frame of reference for evaluations. The results logic serves as the basis for evaluation in order to substantiate results (before, during, in the run-up to or after completion of a project).

The results model is characterised by the following **features**:

- It describes the **changes (results)** that are causally related. Positive changes can affect each other. This means that mutual feedback loops may arise that sustainably reinforce change processes.

- The results model presents an intended **change process** in the area of social concern. Since it is drafted with all key actors, the results model ensures that the project is compatible with both the structures and the processes in the area of social concern.

- The changes that activities aim to influence are only a segment of the change process as a whole within an area of social concern. This segment denotes the **radius of action** within which the project activities are steered and implemented. Other actors may be active in further segments of the area of social concern.

- The full picture that emerges in the results model identifies **alternative options for action** that need to be discussed and decided on through dialogue.

- **Activities** to be implemented are identified for leveraging change. Activities help achieve different results. Through the agreed activities, actors influence changes, and thus the achievement of objectives.

- Steps of change located **outside the radius of action** are less susceptible to influence, but are in the interests of the project. The actors in the project therefore keep this in view and monitor these changes, making the general conditions and risks more clearly visible. These risks can be responded to by selecting flexible activities.

How to proceed

The following sections describe how to devise a results model in five steps.

The sequence of steps described below facilitates the process, but is by no means mandatory. It depicts themes and issues that are interlinked with each other. The steps are therefore designed as reflexive loops. It is thus possible that the participants may subsequently review issues that have already been discussed in previous steps or pre-empt imminent issues.

Step 1: Establish who or what needs to change so that goals can be achieved

To identify the required changes, the cooperation partners should analyse the area of social concern that the project wishes to influence. Discuss weaknesses, influencing factors and potentials. In this context, make use of the outcomes produced by the strategy development process, where available (e. g. process map, map of actors, SWOT analysis, strategic options).

Then build and map out the (intended) change process in the area of social concern. Take a 'top-down' approach, i. e. the starting point is usually an intended overarching result (e. g. derived from the development strategy or planning for the area of social concern, or from a sector strategy). The following key questions will come in useful when designing the change process:

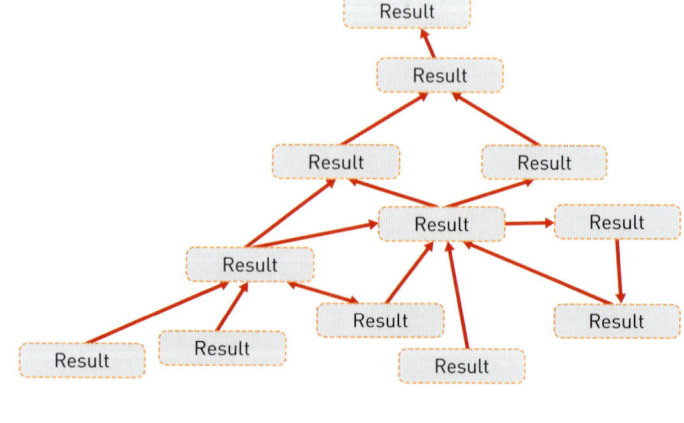

Figure 22: Results model – Step 1

- What has to change? In other words, what must be transformed or needs to improve?

- Which actors have to change their behaviour in order to bring about an envisioned state of affairs?

Formulate necessary change in positive terms. Try to avoid using the word 'not'. Place these identified changes in relation to each other. This provides a systemic, non-linear framework.

The results model homes in on the 'big picture' and on interlinkages, not on the details. It should therefore have a well-defined scope.

Step 2: Identify a joint objective for the cooperation partners

You can turn any result into an objective within a results model. What is important is that the cooperation partners agree on a realistic project objective that can be achieved within a specific time frame.

On the basis of the agreed strategy you define and choose an objective. The results model identifies alternative options for action that you need to review before deciding on one.

The following questions will provide you with a guideline:

- What options for action can be derived from the results model?

- Where does a need exist?

- In which areas do other actors already work?

- Which reasons and interests speak in favour of the individual options for action?

- Are the various options for action realistic?

- On which result or objective would the participants and key actors like to focus? Which objective do all of the participants consider to be a priority?

- What resources are available in the project?

- Can the objectives be realistically achieved with the resources provided by the cooperation partners?

- Is the defined objective realistic?

Define a result as the objective which can be realistically achieved under the given conditions.

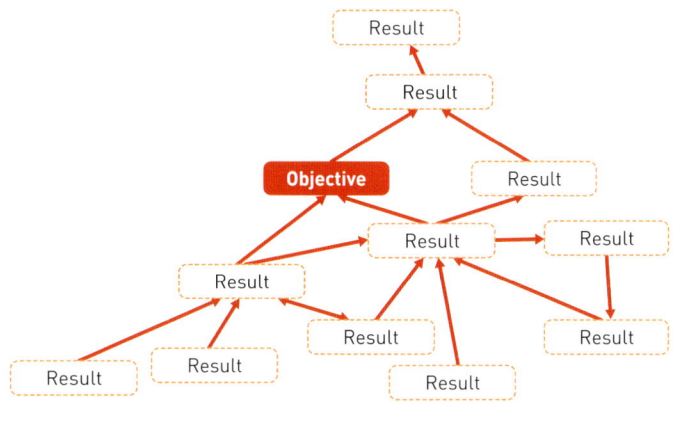

Figure 23: Results model – Step 2

Step 3: Define the internal and external key actors in the project

The results model and a flexible definition of the intended results will help you draft a map of actors– if you have not already drafted one during the strategy process. They will help you answer the following questions:

- Who should be involved in the project? And in what role should they be involved in order to bring about the intended results?

- Who else is working in this area of social concern? Who do we need to keep in mind?

- Within which framework is the project being steered and implemented?

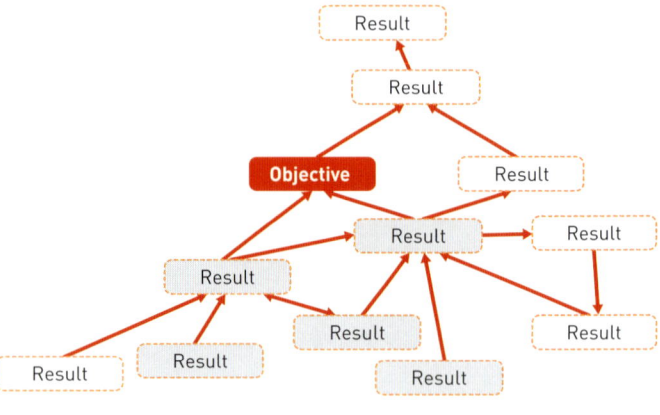

Figure 24: Results model – Step 3

This step is in line with the motto for the success factor cooperation: 'Connect people and organisations to facilitate change.' This is important for broadening the perspective. Besides the cooperation partners who have already been involved, there may be other actors you have not thought of so far who are relevant to the success of the project, and who should therefore be considered and possibly involved in it.

Step 4: Identify the radius of action for which the project assumes responsibility

This is where you define the system boundaries, i.e. the project's sphere of responsibility. The other cooperation partners agree on this sphere of responsibility and in doing so, fine-tune the selected strategy.

The following key questions will help you clarify the radius of action:

- Who needs to be included in project steering, given the sphere of responsibility?

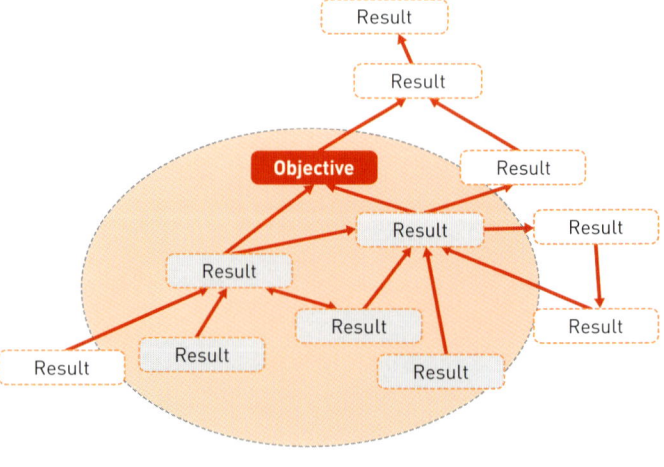

Figure 25: Results model – Step 4

■ Who needs to be kept in mind outside the sphere of responsibility?

■ Which key actors are not yet being involved in the process of devising a results model? How can they be involved and mandated accordingly?

When carrying out the subsequent steps, remember to continuously review whether the sphere of responsibility has been correctly defined or needs to be adapted.

Step 5: Define and agree on the contributions to be made by the cooperation partners in order to achieve results

The individual cooperation partners provide contributions within the scope of activities in order to achieve the intended results.

The following key questions will help in this context:

■ What leverage points should joint activities focus on in order to achieve the intended results?

■ What activities are appropriate and necessary?

■ What resources are available? Are these being put to the most efficient use?

■ Which cooperation partners are responsible for which activities?

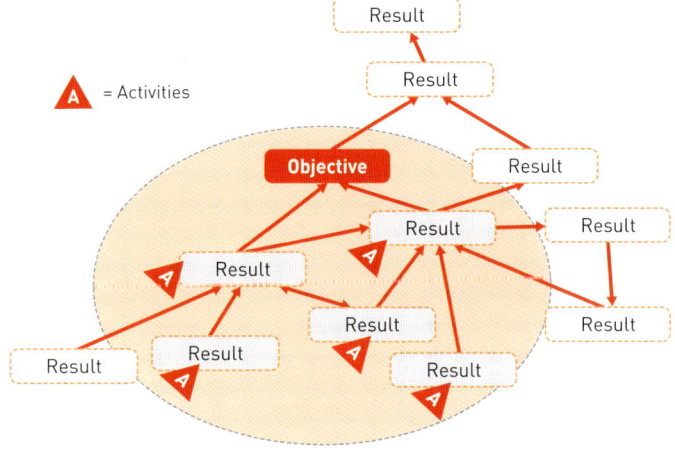

Figure 26: Results model

A rough outline of the activities is depicted in the results model. The area addressed by the activities is of greater importance here.

The activities are depicted in detail at the start of implementation during operational planning (in the plan of operations).

Tool 08
Capacity Development Strategy

Notes on use

Purpose	You can use this tool[11] to review and optimise the activities that are designed to realise a specific strategy at the three levels of capacity development (CD). It will help you to identify interactions between these activities and between the societal (including the elements of cooperation systems and the development of enabling frameworks), organisational and individual levels of CD, and to harness synergies.
When to use it	In all projects that support CD. Successful CD is key to achieving sustainable results in a given area of social concern.
Setting	Workshop with key actors
Facilities and materials	Pinboard and flip chart, workshop materials (markers, cards, etc.), if available: project strategy (e.g. results model), document handouts.
Notes	Before you start working with this tool, a strategic process must already be under way. By the time a CD strategy is elaborated, the (interim) results of this process (e.g. SWOT analysis, description of the selected strategic option, activities carried out by other actors in the sector, etc.) should already be available. The results model elaborated in conjunction with the selected strategic option for project implementation will provide you with a sound basis for designing a CD strategy.

Description

CD should be viewed as a holistic process. In this context, 'capacity' means the ability of people, organisations and societies to manage their own sustainable development processes and adapt to changing circumstances. This includes recognising obstacles to development, designing strategies to tackle them, and then successfully implementing these. This proactive management capacity encompasses the political will, interests, knowledge, values and financial resources that the agents concerned need in order to achieve their own development goals.

The targeted support of CD processes requires a strategy that is geared toward the given political, economic and social context of the area of social concern in question. CD activities must be agreed on with all the relevant actors in the project to ensure that all of them assume ownership of the strategy's implementation. The CD strategy is based on the project's objectives system.

Good CD strategies meet the following **quality criteria**. They:

■ are **embedded** in the context of the area of social concern;

■ are **appropriate** with regard to the actors' willingness to embrace change;

■ are **tied** to existing initiatives in the area of social concern;

■ **coherently interlink** inputs and the results they achieve at the different levels of CD (society, organisation, individual).

The CD strategy defines the specific CD activities within the framework of the strategic option that is to be implemented in the project.

The following table describes the different levels of CD (individuals, organisations, society) along with the potential actors and the methodological approaches and activities involved. The societal level is further broken down into the elements of cooperation systems and enabling frameworks, as this has proven useful when developing activities.

The levels of CD	Actors	Methodological approaches or activities
Individuals **Competence building** *Purpose:* promote personal, social, technical, managerial, methodological and leadership competences in order to develop comprehensive proactive capacities of individuals and networking through joint learning processes	Individuals and communities of learning	Continuing professional development (CPD), training, coaching and knowledge-sharing: improving personal performance and the professional competence of experts, building the creative potential of managers, leaders, change agents and individuals responsible for processes and developing the competence of trainers and consultants in their role as disseminators; networking of individuals for joint, sustainable learning, knowledge creation and dialogue
Organisations **Organisational development** *Purpose:* promote organisational learning and raise the performance and flexibility of an organisation.	Organisations and units of organisations of the state, civil society and the private sector	Change management regarding: agreement on vision and system boundaries, strategy development, strengthening of self-monitoring and learning by organisations, design and start-up of organisations, continuous development of organisations, strengthening of the management system including internal rules and structures, marketing, customer orientation, process optimisation (e.g. of the output processes), HRD systems, project management, finances and other resources, knowledge management.

➡

Society **Development of cooperation partnerships** *Purpose:* establish and develop cooperation between organisations to improve coordination and performance; establish and develop networks for knowledge sharing and co-creation	Institutions and organisations structured along geographical or thematic lines, networks	Relationships and cooperation systems: establishment, development and steering of cooperation systems and networks (e.g. municipal, public-private, sectoral, transnational, product-based) to utilise or capitalise on particular advantages of specific regions and locations and effects of scale, improvement of cooperation relationships to raise performance in the policy field; e.g. the development of sector-specific CPD/education capacities
Society **Development of enabling frameworks** *Purpose:* develop enabling legal, political and socioeconomic frameworks so that individuals, organisations and societies can develop and raise their performance capability	Institutions and organisations (state, civil society, private sector) involved in developing and negotiating the rules of the frameworks concerned	Policy advice: culture of negotiation, opportunities for participation by institutions and organisations, incentives for agreements, agenda analyses, round tables and other forms of participation in the negotiation of rules, interests, basic rights, policies and their implementation, rule of law, checks and balances of power, transparency, mediation and process management of negotiations

Figure 27: Description of the levels of capacity development

Interaction between the different levels

A sound CD strategy will not only generate inputs at the different levels of CD, it will also link these up to form coherent and holistic effects.

Our many years of experience in projects clearly indicate that it only makes sense to focus on a single level if the other levels are being addressed in some other way (e.g. by other actors), or at least are not being ignored.

The following matrix shows examples of the deficits (i.e. imbalances and risks) that can arise from neglecting one level. In turn, these deficits point to opportunities that arise from interaction between activities at the different levels. A strategic input at a particular level will often generate effects at the other levels.

N.B.: It is helpful to illustrate how these deficits might theoretically be manifested in a specific project context, giving examples.

The three levels of CD				
	Activities on the different levels of CD			
Neglect of ... ... affects ▶	**Individuals** Competence development	**Organisations** Organisational development	**Society**	
			Development of cooperation systems	Development of enabling frameworks
Individuals: Competence development		Competence deficit: The individual skills required to initiate and sustainably implement sectoral and organisational change are lacking. Multipliers lack dissemination skills.	Negotiation deficit: The skills required for horizontal cooperation are lacking. Trust deficit: Commitment and personal relationships are underdeveloped.	Leadership and empowerment deficit: Actors do not articulate their interests in negotiations. They are unable to drive strategy development.
Organisation: Organisational development	Transfer deficit: Individuals cannot (fully) apply the lessons they have learned individually within the organisational context.		Deficit of rules: Internal structures and processes are not defined.	Continuity deficit: There is a lack of agreement on rules and process management.
Society: Development of cooperation systems	Knowledge deficit: Horizontal knowledge sharing and continued learning are neglected.	Alliance deficit: Cooperation potentials remain unutilised; one-off solutions and assumed autarchy prevail.		Cooperation deficits: Lack of clarity concerning roles and cooperation between various actors.
Society: Development of enabling frameworks	Deficit of broad-based impact: The skills acquired cannot be sufficiently incorporated into political dialogue.	Frameworks are lacking or are inappropriate: Developed potential goes untapped.	Reliability deficit: Cooperation arrangements and networks remain unstable.	

(left margin, vertical text: Risks of neglecting one aspect)

Figure 28: Interaction between CD activities

How to proceed

Step 1: Understand the concept of a CD strategy

Before work starts on the CD strategy itself, you should ensure that all those involved have the same basic understanding of what a CD strategy is. You can use Figure 27 as a basis for reaching this joint understanding.

Step 2: Determine the focus of the CD strategy

In this step, you define the focus of the CD strategy (the project as a whole, individual lines of action, etc.). In complex projects, it will help if you compile a CD matrix for different parts of the project, so that you don't overload the tool.

CD matrix **Focus:** *(Overall project, line of action, etc.)*				
Phase **from xxx to xxx**	**Individuals**	**Organisations**	**Society**	
	Competence development	**Organisational development**	**Development of cooperation systems**	**Development of enabling frameworks**
Strengths, weaknesses, opportunities, threats (SWOT) in the area of social concern				
Intended capacities				
Activities and hypotheses				
Interaction with the other levels				
Complementary activities by other projects/actors in the same line of action				

Working aid 7: Capacity development (CD) matrix

Step 3: Define current and intended capacities

The first two lines of the capacity development matrix shown above provide a structure for this step. The first line contains the relevant products of the strategy development process (analysis of the actual situation in the area of social concern, current capacities, i.e. strengths, weaknesses, opportunities and threats at the different levels of CD) for the selected focus.

The second contains the intended capacities for each level of CD. Here, the results model is taken as a starting point. The results described there are used to work out the intended capacities at each level. The intended capacities may already have been defined during strategy development.

If not, the key question here is: What capacities, knowledge, political will and other prerequisites are vital if the project is to be a success? What changes will have been brought about?

You often need to further break down the results in the results model, particularly if various activities are required at the different levels of CD in order to achieve one result.

Step 4: Devise activities and hypotheses

You also use the results model to devise activities and hypotheses: What activities and outputs does the project support to underpin the CD process in the area of social concern? What are the underlying hypotheses?

The following questions can help you to develop useful activities:

- What can be done in the project in order to further develop or maintain the strengths at the various levels?

- What can be done in order to make use of the opportunities?

- What can be done in order to neutralise the weaknesses?

- What can be done in order to avoid or address the threats involved?

- What effects can inputs at one particular level have on the other two levels?

Remember that the effects of the joint project complement the activities carried out by other projects/actors in the same area of social concern. This is taken into account in a separate step (step 6).

Step 5: Discuss the interactions between the three levels of CD

In this step, you discuss the matrix and the examples of deficits (Figure 28) against the backdrop of the findings to date.

In the course of this discussion, it may become clear that above and beyond the planned activities, additional interventions will be required in order to ensure that activities are coherent and mutually reinforcing. You should make use of the opportunity to carry out additional activities at other levels in order to improve effectiveness as a whole, thereby safeguarding the sustainability of results.

If further action is required, you should develop corresponding supplementary activities. It is important that you compare the outcome of this discussion with the results model and that any corresponding adjustments required are made there. You should document the outcome in the corresponding line.

Step 6: Discuss complementary activities by other projects/actors

Activities at the different levels of CD carried out by other actors in the area of social concern (e. g. by national change initiatives as part of other projects) are contributions that could supplement the project's activities. You should present these activities here and discuss their possible complementarity. If it becomes clear that activities are missing at certain levels of capacity development and that the project will not be able to carry them out, you should sound out the potential of other projects/actors to fill these gaps.

Success factor Cooperation

Tool 09
Map of Actors

Notes on use

Purpose	This tool will help you identify and visualise the relevant project actors and their relationships.[12]
When to use it	In situations in which it is important to obtain a picture of the actors involved; monitoring of the relationships among actors over time.
Setting	Groups of different sizes. If the group is large, it is advisable to work in smaller groups.
Facilities and materials	Pinboards, flip charts, workshop materials (markers, cards, pins etc.); possibly pre-prepared table on pinboard.
Notes	'Actor analysis' or 'stakeholder mapping' are other common terms used to refer to the 'map of actors'. It is crucial to begin with a clearly defined issue. The map is a snapshot of the situation at a particular point in time. Actors and their relationships change over time, as does the situation. The map of actors is a key starting point for many other planning and consultancy steps and may be useful at various points throughout the life of the project. Using this tool will help you lay the groundwork for using other tools, particularly the steering structure tool.

Description

Actors who hold at least a potential stake in the changes to be brought about by a project, for example, are also referred to as stakeholders. The material resources, social position and knowledge of these actors make them particularly potent, which enables them to wield significant influence over the design, planning and implementation of a project.

Depending on the issue at stake, actors will be either more or less relevant and influential. Drawing up a map of actors for a specific issue means visualising all actors according to their roles and relevance.. You can distinguish between 'primary actors', 'secondary actors', 'key actors' and 'veto players', whereby the boundaries between these categories are usually fluid. **Primary actors** are those actors who are directly affected by the project, either as the designated beneficiaries, or because they stand to gain – or lose – power and privilege as a result of the project. This category includes those who are negatively affected by the project. **Secondary actors** are those actors whose involvement in the project is only indirect or temporary, as is the case for instance with service providers.

Actors who are able to use their skills, knowledge or position of power to significantly influence a project are termed **key actors**. They are usually involved in making decisions within a project. Actors without whose support and participation the targeted results of a project cannot be achieved, or who may even be able to veto the project are termed **veto players**. Veto players can be key, primary or secondary actors. The stronger and more influential an actor is, the more this actor will tend to see himself or herself as the sole actor, and may seek to speak on behalf of or exclude other actors. In other words, in the process of negotiating participation, actors position themselves not only through their relationship to the issues at stake, their institutional position or their resources, but also with respect to the power they have to influence the participation of other actors.

You produce a map of actors by identifying and visualising the (type of) relationships between the actors involved in a cooperation system. The map provides an overview of the entire range of actors involved in the system, allowing you to draw conclusions and formulate hypotheses on the actors' influence on issues addressed in the project and its change objectives, and concerning the actors' mutual relationships, power constellations and dependencies. The roles played by the different actors (primary, secondary, key actors) depend on the specific issue to be addressed. The map offers insights into actual and potential alliances and conflicts. Discussing the map of actors can help you to formulate strategic options and hypotheses concerning specific actors.

The map of actors usually also exposes information gaps and participation deficits (blank spots). It shows the actors and relationships between actors you know too little or nothing at all about, where you need to obtain further information, and which actors you need to involve in the project. The map of actors also corrects premature assumptions concerning individual actors and the relationships between them. Seen in the context of other actors, supposedly important actors become less significant, and apparently insignificant actors take centre stage.

To prepare an accurate map of actors you need to:

Define and demarcate the scope
Start by clearly formulating the key issue in order to circumscribe the area to be mapped and clearly determine the number of actors to be included.

Define the point in time and intervals
The actors form a dynamic system of mutual interdependencies. This web of relationships can change very quickly. It is therefore important that you note the point in time at which the analysis of these relationships was carried out.

Separate the perspectives
Each actor has his or her own perspective. A map of actors therefore only ever represents the perspective of the individuals or groups involved in preparing it.

Key questions for the map of actors

- What do you want to achieve using the map of actors? What specific issue do you wish to address?

- When do you draw up the map of actors and when do you update it?

- Whom do you wish to involve in drawing up the map of actors?
- Were maps of actors drawn up for an earlier phase of the project? You may wish to use them for comparison purposes.

How to proceed

Step 1: Formulate the key issue

By producing a map of actors, what issue do you wish to address at a specific stage of a (future) project? The answer will assist you in steering. It is a good idea to write down this issue on a flip chart so that it is visible while you are working through it.

Step 2: Identify the actors

First of all, identify all the actors relevant to the project or a specific issue. Then assign each of them to one of three groups, namely key actors, primary actors and secondary actors.

To create a map that will yield useful information remember to include all the main actors, without overloading it with too many visualised elements.

Step 3: Select the form of representation

You can visualise the map of actors in two forms, as an onion or as a rainbow.

Both options allow scope for assigning the actors to one of the following three sectors: the state (public sector), civil society or the private sector (you may need to differentiate between other sectors in specific cases).

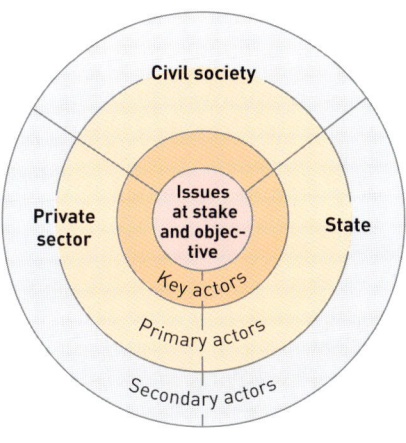

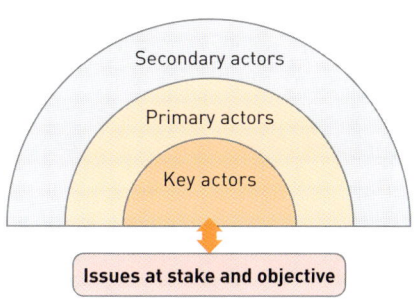

Working aid 8: Map of actors – the onion   Working aid 9: Map of actors – the rainbow

Step 4: Put in the actors

We recommend that you use the same symbol, for example a circle, to represent key actors and primary actors (both of which directly influence the project). The size of the circle represents the actor's influence with respect to the issues at stake and the change objective. Use the letter 'V' to indicate if an actor is a veto player and a rectangle to represent a secondary actor (actors that are not directly involved but may nevertheless exert influence).

●	Key or primary actor with little influence
●	Key or primary actor with little influence
V	Veto player
▭	Veto player

Figure 29: Symbols for actors

You can now position the individual actors accordingly against the selected background (onion or rainbow). It is helpful if you position actors between whom a close relationship exists close to each other. The distance between actors will then indicate how close their relationship is.

Step 5: Represent the relationships between actors

In this step, you show the relationships between the actors. We recommend that you use a different symbol to represent the different type and quality of relationship.

———————	Solid lines symbolise close relationships in terms of information exchange, frequency of contact, overlap of interests, coordination, mutual trust, etc.
- - - ? - - -	Dotted lines symbolise weak or informal relationships. The question mark is added where the nature of the relationship is not yet clear.
════════	Double lines symbolise alliances and cooperation partnerships that are formalised contractually or institutionally.
——————▶	Arrows symbolise the dominance of one actor over another.
——⚡——	Lines crossed by a bolt of lightning symbolise relationships marked by tension, conflicting interests or other forms of conflict.
——╫——	Cross lines symbolise relationships that have been interrupted or damaged.

Figure 30: Symbols for visualising the relationships between actors

Depending on whether you have used the onion or the rainbow, your map of actors will look like one of the two examples shown below:

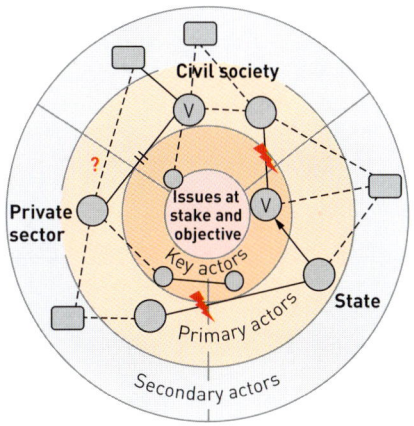

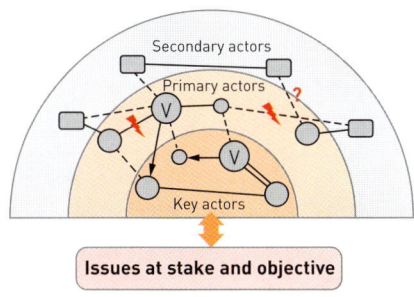

Figure 31: Example of map of actors in
the onion format

Figure 32: Example of map of actors in
the rainbow format

Step 6: Evaluate the outcome

In this last step, you jointly evaluate the outcome. Is your map of actors an accurate reflection of the current reality? Have you taken into account all relevant actors? What is the first thing that strikes you? What do you think of the picture the map gives you? Are any important elements missing?

We recommend that you use a flip chart to document the key outcomes of your joint discussions. This should include working hypotheses and possible options for action, presented in relation to the issue defined at the outset (see step 1).

Tool 10
Actor Profiling (4 A's matrix)

Notes on use

Purpose	This tool is designed to help you identify the attitudes of relevant actors toward the project. Groupings are brought to light, such as supporters and opponents who may block progress. The actor profiles provide you with a basis for discussing and comparing strategic options.[13]
When to use it	In situations where you need to identify the attitudes of the actors relevant to the project. This tool will come in useful in an actor analysis (together with the 'Map of actors' tool, for example).
Setting	Up to 25 participants, preferably in small groups, group rooms may be required
Facilities and materials	Pinboards, workshop materials (markers, cards etc.); handouts of the evaluation criteria; pre-prepared matrix of actors on pinboard (see working aid 10 below).
Notes	▪ You will require relatively accurate knowledge of the actors to be evaluated. ▪ By its very nature, evaluation is a subjective judgement.

Description

The strategic orientation of a project is the result of a negotiation process. To identify and compare the various perspectives and interests of the participating actors, it is helpful to ask the following questions:

▪ What agenda do the actors have? In other words, what mandate, strategic objectives and interests do they have?

▪ In which arena or area of activity do they act? How great is their scope of influence?

▪ What alliances have they formed with other actors?

Drawing up actor profiles based on various criteria will help you visualise the relative importance of actors and decide whether relationships between actors need to be established and developed. They will also help you identify potential groups of actors that share similar profiles. Groups of this kind are important in change management, because actors with similar profiles can reinforce each other's supportive or critical attitudes toward the change objective.

How to proceed

Step 1: Identify actors

In a first step, list the actors who are connected with the theme and with the change objective. A 4 A's matrix (actors, agenda, arena and alliances) clarifies where information gaps need to be closed, and provides you with an initial overview.

Issues at stake and change objective:			
Actors name, core function	**Agenda** mandate/mission, strategic goals	**Arena** area of activity, scope of influence	**Alliances** relationships to other actors
Actor 1			
Actor 2			
Actor n			

Working aid 10: 4 A's matrix

Depending on the level of detail required, it may be helpful to further differentiate alliances in terms of quality, for example:

- A institutional dependency
- B continuous exchange of information
- C coordinated action
- D co-production using shared resources
- …

You can also use the matrix as a periodic monitoring tool to track changes in the map of actors over time.

Step 2: Transfer the results into an actor profile

In this step, you transfer the findings of the 4 A's matrix for the key actors into the following working aid. You can, of course, modify or add to the ten items listed below. Once you have finished, amalgamate your different ratings to determine the corresponding actor profile.

Items	--	-	+	++
Vision of development: The actor possesses a constructive vision of development based on democracy and fairness.				
Managing for results: The actor acts in a results-based manner and periodically reviews the achievement of results.				
Flexibility and innovation: The actor is open to new ideas and adjusts his/her organisation to new challenges.				
Contractual loyalty: The actor sticks to agreements and meets their provisions on time.				
Communication: The actor actively informs partners of his/her activities, exchanges information and answers enquiries swiftly.				
Relationships: The actor facilitates contacts, creates spaces for encounter and adjusts his/her actions to the capabilities of his/her external partners.				
Management: The actor acts on the basis of transparent guidelines and strategies as well as defined roles and responsibilities.				
Trust: The actor informs others proactively of his/her intentions, aims and expectations, and shows understanding for others' interests.				
Conflicts: The actor draws attention to tensions early on, and is willing to address them constructively, openly and quickly.				
Capitalising on experience: The actor evaluates his/her experiences, is open to criticism and shows a willingness to learn and to change.				
-- / ++ = agree/disagree				

Working aid 11: Actor profile

Step 3: Compare actors and determine implications

Comparing the profiles of the various actors is not only an instructive exercise, it will also help you identify socio-cultural behavioural patterns (e.g. clientelism, authoritarianism, religious orientations) that can play a major role both in public administration and in private-sector organisations.

Step 4: Compare with strategic options

Actor profiles provide you with a sound basis for discussing potential profiles for action and thus for determining key challenges facing the different actors. Identifying these challenges will help you to talk through and compare different strategic options.

Tool 11

Interests of Key Actors

Notes on use

Purpose	This tool will help you identify key actors and their interests in the project's change objective.[14]
When to use it	Well suited to coaching of or self-reflection by responsible actors or to reflective work within a core team for formulating hypotheses.
Setting	Workshop
Facilities and materials	Document handouts
Notes	Before you start using this tool, participants must have a good knowledge of the key actors and be able to assess them.
	As the tool evaluates the interests of key actors, a high degree of openness and trust will be required if it is to be used jointly with these key actors.
	Ideally, you will use this tool once you have completed the map of actors.

Description

This tool will help you shed light on the different interests that the most important actors have in the change objective. You must therefore have already identified these actors (for example, by completing a map of actors or a 4 A's matrix). Key actors are characterised by the following aspects:

Legitimacy: Institutional position of the key actor, ascribed or acquired rights that are for instance underpinned by the law, an institutional mandate and public approval, and are considered legitimate. This also includes key actors without whose explicit approval the project would be inconceivable. These veto players can create key impetus and scope, or they can obstruct the project.

Resources: Knowledge, expertise, skills and material resources that enable the key actor to significantly influence the change objective, or to steer and control access to these resources.

Networks: Number and strength of relationships with other actors who are obligated to or dependent on the key actors. Key actors are usually well networked, i.e. they have a large number of institutionally formalised and informal relationships with other actors. Key actors therefore wield significant influence on the participation of other actors, shaping some decisions as to whether certain actors will be included or excluded.

The interests of the key actors are usually not fully congruent with the change objective. This is only natural, given that a project by its very nature is usually of an innovative nature. Any change will also generate responses of reserve and resistance. The actors notice the dissonance between their interests and the change objective at the latest when they are called upon to depart from familiar paths and learn new approaches. This can create tacit or explicit resistance in various forms: reserve, sceptical aloofness, objection or openly organised resistance to the targeted changes.

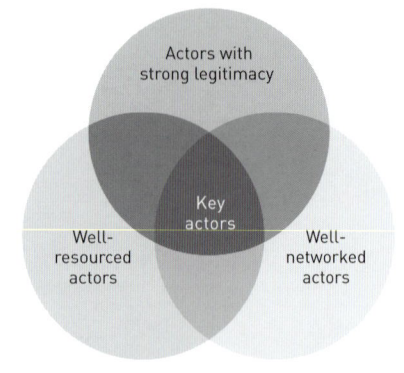

Figure 33: Identifying key actors

The project must address this resistance. In order to do so, this resistance must be clearly articulated. There are many possible motives for resistance and these are closely linked to the change management process, for example through actors' self-interest and fear of losing power or their mistrust of other actors. Unclear information about the project also reinforces resistance. If resistance remains based on (tacit) assumptions or speculation, because it cannot be expressed or is not taken seriously, then it will also increase. And what begins as verbal assent may in the course of the project turn into reserve or even resistance.

To prevent a desired project from being vetoed, it is vital that the interests of the actors are understood. Once the key actors have articulated their perspective it is possible to alleviate feelings of uncertainty and address the resistance early on, so that a negotiation-oriented open climate for achieving the desired changes can be created.

How to proceed

Step 1: Establish degree of compliance with the change objective

When analysing the attitudes of the key actors to the change objective, it is important that you ask the following questions:

- What interests do the key actors have in the change objective?

- To what extent do these interests comply with the change objective?

- What effects does this compliance/lack of compliance have on the change objective?

- What strategic options do you need to develop to broaden the scope for action, win the support of actors and eliminate obstacles (for example, in relation to information and communication, structuring participation, strengthening relationships between actors, improving access to new knowledge and supporting negotiation processes)?
  How should you manage the change process so that the key actors can be involved effectively?

You should discuss these four dimensions for each relevant key actor, and summarise the findings in the table below.

Issues at stake and change objective of the project:				
Key actors	Interests In issues at stake and change objective	Compliance with the change objective from -- to ++	Possible effects of harmony/ dissonance/ indifference	What to do? Options for broadening the scope for action
Actor 1				
Actor 2				
Actor n				

Working aid 12: Compliance with the project's change objective

Step 2: Map conflicting objectives

Matrix of conflicting objectives			
Key actors	Compliance with the change objective from -- to ++	Change in terms of: ■ legitimacy ■ resources ■ networking	Fears and anticipated losses
Actor 1			
Actor 2			
Actor n			

Working aid 13: Matrix of conflicting objectives

In order to shed a more precise light on any objectives that may conflict with the project, you now establish the degree to which the vision of each actor corresponds with the change objective. To do this, answer the following questions:

■ To what extent might the project change the legitimacy, access to resources and networks of the key actors?

■ What fears or anticipated losses might motivate the actions of the key actors?

Step 3: Discuss the matrix of conflicting objectives

In this step, jointly discussing the situation as captured by the simplified matrix on conflicting objectives can:

- help identify beliefs that actors share. For instance, actors in the central government administration may fear that they could lose legitimacy and influence in a decentralisation process.
- enable planners to address and work through conflicting objectives with the key actors early on. In the case of a decentralisation process this could mean, for instance, broadening their mandate to include new tasks of regulation, supervision and support of municipalities.

The varying degrees of compliance with the change objective affect the project, and wherever possible should be taken into account early on when devising and selecting strategic options.

Step 4: Address conflicts

More often than not, projects usually also need to deal with **conflict in relationships and clashing interests** among the actors. The first question you need to ask is whether the conflict should be made explicit and addressed at all. This question is crucial, because addressing tension and conflict always has positive spin-offs/results. For example, it could help clear up other unresolved issues, such as the division of roles between the actors.

Any conflict of relationships and interests comprises **three basic elements**: the two parties to the conflict, and the issue at stake that is giving rise to the conflict. The two parties to the conflict normally hold contrary positions. Each party is annoyed by the other and attempts to weaken the other's position and to strengthen their own. If we wish to address conflict, we need to place actors in relation to each other in order to transform these positions into different interests. This takes place in three phases, as illustrated below:

Phase 1:
We hold contrary positions. The other actor is the problem, he/she is inflexible and stubborn. We stick to our position, because we're right.

Phase 2:
We focus on the issue at stake. We see the issue differently, and we recognise the fact that our interests are different.

Phase 3:
We study the issue in greater depth. We find that exchanging different perspectives and negotiating interests leads to a compromise or a viable agreement.

Figure 34: Phases in dealing with conflict in relationships and clashing interests

Tool 12
Structural Characteristics of Cooperation

Notes on use

Purpose	This tool lists eleven key structural characteristics that will enable you to reflect on the quality of cooperation. These characteristics will come in useful when discussing alternative forms of cooperation with others and identifying specific activities. The structural characteristics home in on the relationships between the cooperation partners rather than on the traits of the individual actors themselves.
When to use it	To define, for example, a temporary cooperation system's boundaries during the initial phase of a project and to reflect on the quality of the cooperation at routine intervals.
Setting	Small group of between two and ten people.
Facilities and materials	Handout: list of structural characteristics (self-check) (working aid 14); pinboard and flip chart, workshop materials (markers, cards, etc.), if available: map of actors.
Notes	A sound understanding of the structural characteristics is vital. It is helpful if you have compiled a map of actors before you start using this tool.

Description

Each cooperation system develops specific forms of cooperation, which are epitomised by eleven key structural characteristics.[15] These characteristics can be assigned to four different areas:

- **Structure of the actors involved**

- **Ties between the actors**

- **Quality of communication**

- **Rules and roles**

Structure of the actors involved: The number of actors involved and their diversity influences the form of cooperation used. A cooperation system that only has a small number of actors that are similar to each other is easy to steer, provided the balance of power is relatively stable. The downside, however, is that it will often lack innovation and effectiveness. The more actors are involved and the greater their diversity, the more difficult the system is to steer.

The structural characteristics of this particular aspect are:

- Number of actors

- Heterogeneity of the actors

- Influence exerted by specific actors

Ties between the actors: The objective a project sets out to achieve will determine just how intense cooperation needs to be. Is it a simple matter of exchanging information? Or will the actors have to forge links with each other and pool resources to come up with harmonised strategies (co-production). The intensity of cooperation will vary accordingly. The cooperation term must correspond with the objective of the joint project and the willingness of the actors to cooperate, as must the scope that partners have to leave or allow new partners into the system.[16]

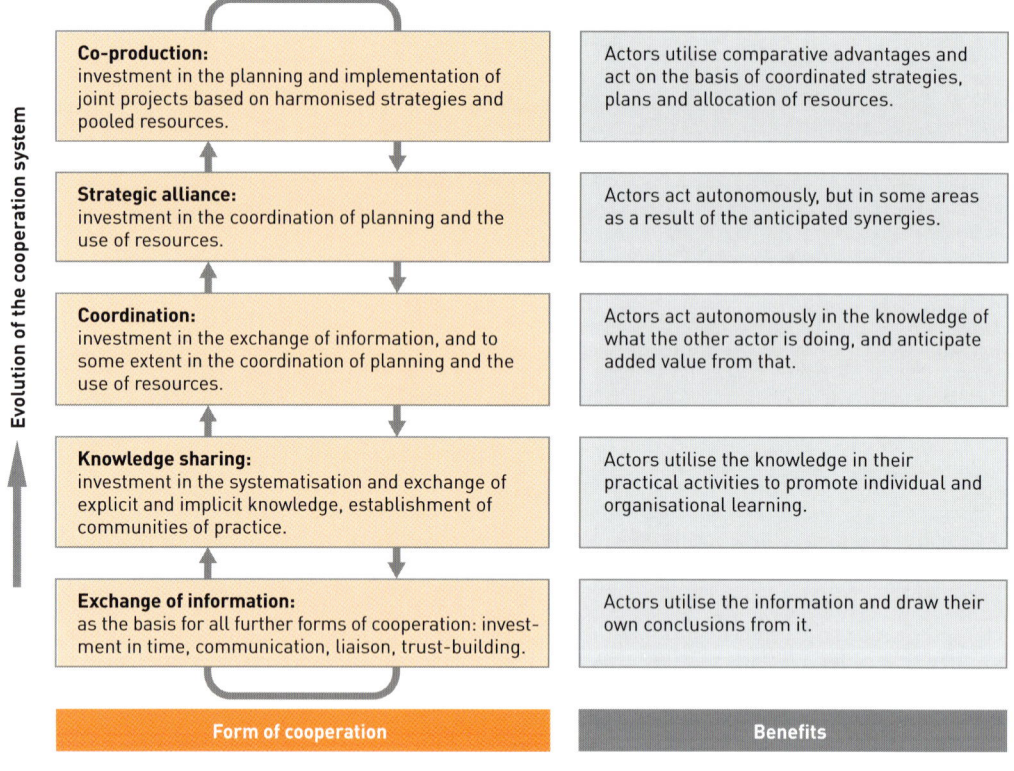

Figure 35: Stages of development in cooperation systems

The structural characteristics of this particular aspect are:

- Openness (flexibility) of the cooperation system
- Length of the commitment
- Intensity of cooperation

Quality of communication: The quality of communication between the cooperation partners is determined to some extent by their personal relationships and the frequency of exchanges. The quality of the 'social spaces' in which communication occurs also plays a key role, however. There are different types of 'social space', including:

- Committees such as 'steering groups', 'monitoring groups' and 'secretariats' are part of the formal *steering structure* (see the success factor steering structure for more information). They are designed for coordination and decision-making in a cooperation system and require a strong commitment.

- 'Focus groups' or 'sounding boards' are social spaces in which *formalised exchange* can take place with those actors that are not directly involved in decision-making. They usually aim to obtain additional information as a basis for making future steering decisions and provide feedback on decisions that have already been made.

- 'Round tables' are convened *for a specific reason*, usually to clarify acute, conflict-related issues.

- 'Working groups' or 'regular operational meetings' are held at *routine intervals* to coordinate and implement operational activities.

- Networks are a special type of social space in that they provide the opportunity of forging relationships with other actors swiftly and efficiently where needed (see the success factor cooperation for more information). 'Networking' involves establishing new contacts – which could blossom into *future relationships* – through events, conferences and direct communication. A project can set up and consolidate networks or tap into existing networks.

- Informal social spaces (such as working lunches, fireside chats) offer a good opportunity for *building trust* and often play a key role prior to making a decision.

The structural characteristics of this particular aspect are:

- personal relationships;

- intensity of coordination;

- quality of the 'social space'

Rules and roles: Successful cooperation partnerships require striking a balance between formality – i.e. introducing rules – and informality – allowing a certain degree of flexibility. Finding the appropriate degree of formalisation is often comparable to walking a tightrope. On the one hand, you must ensure transparency and commitment while on the other, you need to ring-fence the transaction costs involved in establishing rules and roles (risk of an 'inward-looking approach'). Differentiating between roles in the cooperation system will help increase efficiency and minimise conflict. Cooperation partners can choose to step into a number of different roles (cf. Figure 36 below).[17]

Roles	Symbol	Brief description
Node		The node contains the core tasks: point of coordination and communication, networking among the actors, initiation of projects involving several actors.
Manager		As a manager the actor plans and implements, together with other actors, individual projects that are decided upon and supervised to completion by the cooperation system as a whole.
Spokesperson/ advocate		As a spokesperson or advocate the actor represents the interests and concerns of the cooperation system and the projects vis-à-vis the public or politically relevant bodies.
Negotiator		As a negotiator the actor has a mandate to represent and negotiate the concerns of the cooperation system with third parties.
Process management and facilitation		The actor shapes the process architecture, organises the process and facilitates negotiations in the cooperation system.
Consultant		As a consultant the actor contributes knowledge and experience, and promotes self-reflection within the cooperation system. He/she is commissioned by the cooperation system and can also perform coaching functions for other actors.
Connector		As a connector the actor creates links that are important for a certain project, such as links to the institutional environment, e. g. governmental agencies.
Supporter		As a supporter the actor is available for various support activities such as support and consultancy, or back office duties for smaller projects.
Participant		The actor participates in a project, e. g. as a service provider or financing body.
Observer/ feedback provider		The actor observes the activities of others and carefully communicates his/her observations and perceptions. He/she is a professional feedback provider.

Figure 36: Forms of cooperation and roles

The structural characteristics of this particular aspect are:

- Degree of formalisation
- Division of roles

How to proceed

Use this tool to assess the quality of existing or future cooperation systems based on common structural characteristics. You can use it for the entire cooperation system or for sub-systems.

Step 1: Use structural characteristics to carry out self-check

Use the following structural characteristics to assess the quality of the cooperation system. Ask yourself the question: Is the corresponding structural characteristic suitable for the cooperation system's objective and the issues at stake? Is an appropriate design used?

Suitable/appropriate

Suitable/partially appropriate

Unsuitable/inappropriate

A) Structure of the actors involved

Number of actors: As the number of actors increases, the negotiation and steering requirements rise exponentially. Groups of free-riders, thematic satellite groups and power circles form. Is the number of actors involved appropriate for the cooperation system's objective and the issues at stake?	
Heterogeneity of the actors: Homogeneous groups of actors (i.e. ones that are similar in terms of sector, activity areas, size, life cycle, region of origin etc.) usually lack innovation or tend to be competitive. Heterogeneous groups have a lot of potential for innovation, but disintegrate if their diversity is not capitalised upon. Is the heterogeneity of the actors involved appropriate for the cooperation system's objective and the issues at stake?	
Influence exerted by specific actors: An individual or small number of key actors may wield major influence over cooperation. In some cases, they may even dominate. The partners may be largely equal in terms of their degree of influence, however. In terms of the cooperation system's objective and the issues at stake, is there an appropriate weighting as regards the influence exerted by the actors involved?	

B) Ties between the actors

Openness (flexibility) of cooperation: The actors involved may show a varying degree of interest in involving new partners. New partners may not always find it easy to join the cooperation system. Too much openness can overstretch the cooperation system, as it constantly strives to integrate new partners. Too little openness can stifle growth and the capacity to innovate. Is there an appropriate degree of openness in terms of the cooperation system's objective and the issues at stake?	🟢 🟡 🟠
Length of the commitment: A cooperation partnership can run on a short, medium or long-term basis. Temporary cooperation systems (projects) may become institutionalised over the long term (for example, decentralisation as a new system of cooperation between the central state and municipalities), but they can also become moribund if the benefit to the participants is not evident and there is no joint strategic orientation. Is the length of the commitment appropriate for the cooperation system's objective and the issues at stake?	🟢 🟡 🟠
Intensity of cooperation: The five stages of development in cooperation systems (cf. Figure 35) require an increasing degree of mutual obligation and dependability: exchange of information, knowledge transfer, coordination, strategic alliance and co-production The (willingness to form) ties must be in keeping with the cooperation partnership's original ambitions. If it is too weak, then it will be impossible to reach the next stage. Is the intensity of cooperation appropriate for the cooperation system's objective and the issues at stake?	🟢 🟡 🟠

C) Quality of communication

Personal relationships: In the best case scenario, the level of personal relationships is characterised by mutual trust and professional respect. If the quality of the relationship is inadequate (e.g. if there is mistrust or a lack of respect) or if too much emphasis is placed on the personal relationships level (e.g. in the form of cliques) then the achievement of objectives in the cooperation system may be severely jeopardised. Is the quality of the personal relationship appropriate for the cooperation system's objective and the issues at stake?	🟢 🟡 🟠
Intensity of coordination: How necessary is it for actors to see each other and coordinate their activities on a regular basis? The costs of coordination must be proportionate to the anticipated benefits. Generally speaking, cooperation systems need the actors to meet face-to-face. This reduces the number of possible interfaces. Is the intensity of coordination appropriate for the cooperation system's objective and the issues at stake?	🟢 🟡 🟠
Quality of the 'social spaces': Communication can occur bilaterally or at 'round tables', in one-off events for large groups or in regular working groups. It can be written or verbal, and organised on a face-to-face or virtual basis. If the available 'social spaces' are unsuitable, cooperation will become inefficient and lack innovation. Is the quality of the 'social spaces' appropriate for the cooperation system's objective and the issues at stake?	🟢 🟡 🟠

D) Rules and roles

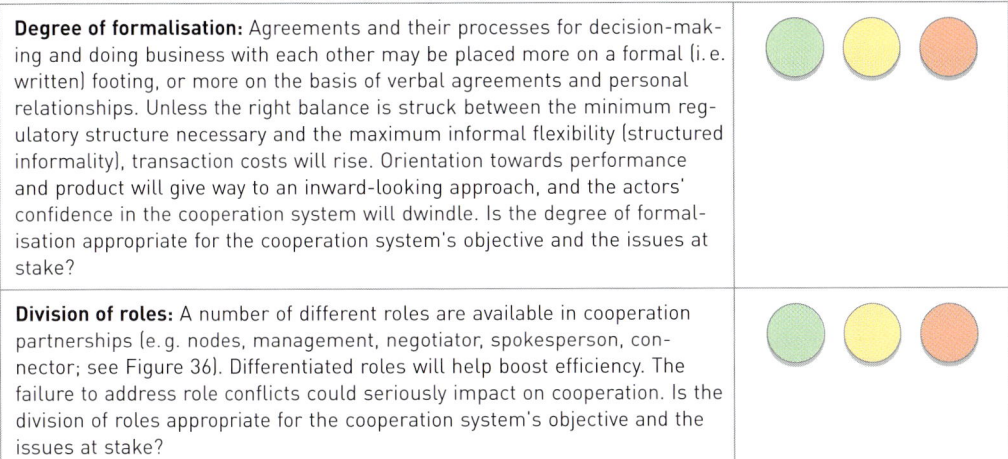

Degree of formalisation: Agreements and their processes for decision-making and doing business with each other may be placed more on a formal (i.e. written) footing, or more on the basis of verbal agreements and personal relationships. Unless the right balance is struck between the minimum regulatory structure necessary and the maximum informal flexibility (structured informality), transaction costs will rise. Orientation towards performance and product will give way to an inward-looking approach, and the actors' confidence in the cooperation system will dwindle. Is the degree of formalisation appropriate for the cooperation system's objective and the issues at stake?	
Division of roles: A number of different roles are available in cooperation partnerships (e.g. nodes, management, negotiator, spokesperson, connector; see Figure 36). Differentiated roles will help boost efficiency. The failure to address role conflicts could seriously impact on cooperation. Is the division of roles appropriate for the cooperation system's objective and the issues at stake?	

Working aid 14: Checklist for structural characteristics of cooperation systems

Step 2: Identify areas for improvement

In this step, you develop activities that will help improve the cooperation system. The following guidelines will help you in this context:

- You can address any deficits in the **structure of the actors involved** by drawing a clear distinction between internal and external cooperation: You can try to involve additional actors in the 'inner circle' of cooperation and to clearly differentiate between 'internal' and 'external' actors using the map of actors.
  You can take structural measures (e.g. delineating between the strategic and the operational level, see the success factor steering structure for more information) and clarify roles in order to better integrate challenging positions of power wielded by individual actors.

- The need to strengthen **ties between actors** indicates that the objective has not been clarified to a sufficient degree or that there are unresolved issues between the actors. In this context, it is helpful to consolidate the joint basis for cooperation either by carrying out trust-building activities (see the tools 'Trust-building' and 'Shaping negotiation processes'), clarifying the objective or by raising awareness of shared history (for example through story-telling, which involves actors describing achievements, heroic deeds and the obstacles overcome during cooperation from their own perspective).

- In many cases, it is only possible to wield indirect influence over the quality of **communication between actors**. Identifying and managing conflict may help in this context, as can tailoring communication platforms to specific needs (adequate time, no interruptions, confidentiality) and strengthening the actor's own cooperation and communication skills.

- The clarification of **rules and roles**, i.e. formalising cooperation to a greater degree is often a relatively simple way of addressing deficits in this regard. A note of caution, however: stipulating too many rules and regulations may prove counterproductive – additional rules will only generate additional value if those already in place are being complied with.

Answer the following questions for the need for action identified in step 1:

- **Assessment:** To what degree does the need for action identified influence the achievement of objectives in the project?

- **Hypotheses:** What could be the underlying cause of existing deficits? As regards the need for action identified, where do specific potentials lie?

- **Identification of possible activities:** What activities would improve cooperation (in terms of the specific need for action)?

- **Evaluation of the proposed activities:** How effective is the activity considered to be? How innovative is the activity in the cooperation system (or is it just 'more of the same')? How realistic is the activity? To what extent are the actors in the cooperation system prepared to engage in the activity?

Step 3: Draw conclusions and make agreements

In this final step, you agree on activities and start implementing them. A few weeks after implementation, you should review and reflect on whether the situation has changed.

Tool 13
Views of Actors (PIANO Matrix)

Notes on use

Purpose	This tool is designed to help you to quickly assess five relevant aspects of cooperation systems from the point of view of the different actors and to define appropriate activities.[18]
When to use it	To address and periodically monitor and evaluate critical situations you encounter when designing cooperation systems. You can add your own key boxes to the PIANO matrix if necessary.
Setting	Workshop with key actors
Facilities and materials	Flip charts, workshop materials, pinboard with pre-prepared PIANO matrix.
Notes	Participants must have a clear understanding of the project objectives and corresponding expectations placed on the network.

Description

Negotiation processes are a defining feature of cooperation systems, which are not hierarchically structured. This requires effective communication among the actors. Objectives-oriented cooperation management requires a clear understanding of the mutual interdependencies between the actors and the interests, incentives and values that drive their actions.

These incentives include in particular:

- economic incentives: access to new resources, market access, access to know-how, use of competitive advantages, leveraging of potential for greater efficiency, for example

- political incentives: power accrual, improving and expanding social relationships, access to information, for example

Cooperation systems function on the basis of successful negotiation and agreement. For cooperation arrangements to succeed, the actors involved must be able to negotiate with each other and make joint decisions. Developing the negotiation and decision-making skills of the actors involved therefore constitutes a key challenge in this context.

It is important that all the actors involved realise that given the interdependencies involved, none of them are able to develop viable solutions on their own. This realisation is by no means a given. After all, the different actors involved in a cooperation system – whether they are from the public sector, civil society or the private sector – usually come from a wide range of communication and decision-making cultures. The more understanding the actors involved have for the differing interests of the other parties, for incentives structures, patterns of communication and for values, the easier it should be to coordinate the different systems of logic. The tool described here will allow you to familiarise yourself with the different perspectives that the actors involved have on the cooperation system. It can be used to address and periodically monitor and evaluate critical situations that may arise in all of the development phases of the cooperation system. The aim here is to strengthen the following elements of a cooperation system:

- functional and active participation of the actors

- strengthening of the shared vision and orientations

- trust-building and consolidation of relationships among the actors

- transparency as regards the different degrees of influence and implementation strategies used by actors

- strengthening of the actors' identification and motivation

- equal access to information

- maximum rate of learning through information exchange

- communication on the cooperation system and securing of recognition

- minimisation of transaction costs for coordination and cooperation management

A cooperation system is perceived and judged differently by the participating actors, depending on their perspective and particular interests. To make these different perspectives visible and negotiable, it is helpful if this tool is applied separately by different groups of actors. If required, each group of actors can therefore carry out the proposed sequence of steps separately.

How to proceed

Step 1: Discuss and fill in the PIANO matrix

In this first step, you fill in and discuss the individual boxes in the PIANO matrix. The basic version of the matrix contains the following five key boxes for steering cooperation systems: Products, Incentives, Actors, Negotiations and Orientation (PIANO). You can add your own key boxes to the PIANO matrix if necessary. You use the matrix to take stock of the situation. As it is relatively easy to use, it can be used as often as necessary.

Start by filling out the Products, Incentives, Actors, Negotiations and Orientation boxes by answering the key questions shown in the columns below and documenting the answers in writing in an identical table on a pinboard.

P Products	I Incentives	A Actors	N Negotiations	O Orientation
Actual situation				
What do you wish to achieve together with the other actors involved? What products or services do you wish to offer? To whom? What intermediate outcome do you wish to achieve in the next step? What contributions will the individual partners make?	What motivates you to participate in the network over the long term? What benefits and added value do you expect to obtain from that? To what extent do your expectations match or not match the anticipated benefits?	What are your strategic goals? To what extent do your interests and objectives overlap?	What ground rules for internal communication and cooperation do you need to apply to cooperation? How should you ensure that agreements are not broken?	What vision do the agreed objectives set out to realise? Do your visions contradict each other? To what extent? Do you have a common vision? What do you regard as the greatest divergences in the near future?
Areas for improvement				
Are the service delivery processes well coordinated? Which actors outside of your cooperation system do you need to involve to provide the planned products/services?	What options are there for increasing actors' motivation for joint cooperation?	Which actors are you dependent on in order to create the anticipated benefit and added value? Does the mix of cooperation partners need to be changed? If so, in what way? Do you require additional expertise (technical, consultancy) to achieve your objectives?	Are you making optimal use of the know-how that the involved actors bring to the table? Are you using suitable communication structures to achieve objectives efficiently and effectively? How satisfied are you with your decision-making structures and decision-making patterns?	What options are there for strengthening the joint focus?
Activities				
...	...	...	...	...

Working aid 15: The PIANO model

Step 2: Identify areas for improvement

Here, you identify areas for improvement together with other actors based on the existing deficits and the potential for synergies in the cooperation system. You then enter these in the individual columns of the PIANO matrix.

Step 3: Devise activities

In this final step, you define and agree the key activities to be carried out during the course of the project and fine tune existing activities, based on the outcome of the previous steps. You identify and define activities by working through the PIANO table column by column. Once you have completed a PIANO table for a cooperation system, we recommend that you refer back to it at appropriate intervals for discussion and review purposes.

Tool 14

Networks: Strengthening Relationship Potentials

Notes on use

Purpose	This tool will help you decide whether activities to establish or strengthen networks should be carried out. It will help you plan specific activities by providing examples and by stipulating criteria for safeguarding the quality of networks.
When to use it	To prepare activities to strengthen networks.
Setting	Between three and twelve participants.
Facilities and materials	Flip chart, document handouts.
Notes	It is important that you have a clear understanding of the project's objectives and of the corresponding expectations placed on the network. This tool allows you to make an initial assessment of whether and how a network can be used to achieve shared objectives.

Description

Social networks bring together individuals and organisations in loose relations of power. In their strictest sense, networks provide **opportunities** for individuals and organisations with similar interests to come together to forge relationships. The more opportunities there are to quickly and easily establish contact with each other when needed and initiate cooperation, the stronger the network[19].

Links between cooperation systems and networks. Although the two terms are often used synonymously, Capacity WORKS draws a clear distinction between cooperation systems and networks (for more information see the success factor cooperation). A cooperation system is normally embedded in different social networks with relationships to actors in politics, administration, the private sector and civil society. The cooperation system may even have emerged directly from one of these networks. Conversely, it may make sense for a cooperation system to consolidate existing networks or set up new ones. Networks can provide support in developing and disseminating innovative ideas. They can encourage key actors to learn from each other and generate new knowledge together and can help reach out to decision-makers and attract new cooperation partners. Establishing and maintaining helpful network contacts – through targeted community building, for instance – is therefore a key task in relationship management within a cooperation system. In this context, it is important to actively shape networks for potential relationships in order to transform them into operational cooperation relationships if the need arises.

The following three steps will help you make a concrete decision on whether activities to establish or strengthen networks should be carried out.

How to proceed

Step 1: Clarify the goal

Establishing and taking specific steps to build networks is usually time-consuming and can be very costly. Success is not always guaranteed. Before you design activities, it is therefore important that you clarify what particular goals you wish to achieve and whether setting up a network is the most suitable way of achieving them.

The following checklist will help you clarify your goals:

Target group					
The target group is clearly defined and comprises a few select individuals and organisations.	1	2	3	4	A broad spectrum of (diverse) actors is to be addressed.
Commitment					
The actors involved are to agree on a binding goal and achieve it in a transparent manner.	1	2	3	4	First and foremost, the actors involved are to be offered the opportunity for mutual exchange and brainstorming of ideas for a particular theme.
Pressure to succeed					
Key actors expect tangible outcomes to be available as soon as possible.	1	2	3	4	The main priority here is to establish contacts. Tangible outcomes in the form of innovations and specific projects are only anticipated in the medium to long term.
Contribution of actors					
Contributions can be requested from actors on a continuous basis to ensure that the goal is achieved.	1	2	3	4	Participation is to be possible with minimum time and effort, preferably on a non-binding basis.

Working aid 16: Checklist for clarifying the goals of networks

Once you have completed the checklist, address the following questions:

What goal has been defined? What goal will setting up a new network or strengthening an existing one achieve? How will it be evident that this goal has been achieved?

What options are there for achieving this goal? Are activities to establish/strengthen a network the most appropriate way of achieving this goal? The higher the checklist score, the more suited the establishment/strengthening of networks is to achieving the goal. A low score indicates that a cooperation system with clear system boundaries is a more appropriate means of attaining your chosen goal. In this case, activities to strengthen networks can be used as a back-up but other Capacity WORKS tools (such as 'Structural characteristics of cooperation partnerships'), are more suitable for shaping the cooperation system.

Step 2: Identify the qualities of a network

Above all, social networks should be efficient (i. e. enable participants to establish contacts swiftly, easily and at low cost) and robust (i. e. resilient against disruption such as bilateral conflict) and should offer added value to those involved. The following checklist will help you develop hypotheses on how efficient and robust an existing network is and its added value.

Criterion	Comments and hypotheses
'Diversity': The network pools different perspectives on common issues, for example by involving actors from the public and private sector.	
'Triangulation': Actors who are directly linked with each other are also linked through the same external parties. This will create a tightly-knit web of relationships that can be used to resolve conflict in bilateral relationships.	
'Everyone knows everybody': The average distance separating the actors involved is short. This distance is measured in 'steps' (A is in contact with B, who is in contact with C, so that A and C are connected by two 'steps').	
'No actors or groups of actors are isolated'. In other words, all actors involved are connected at the very least by a number of different 'steps'.	
'The key actors are directly connected with each other' (avoiding the formation of cliques, which could have a divisive effect).	
'Clear added value for members': The network generates an obvious added value for its members, for instance by identifying new solutions, generating new knowledge, establishing relationships or consolidating existing ones, adding prestige, providing power of interpretation, creating autonomy, ensuring participation in shared products and motivating through reciprocity and equal exchange.	

Working aid 17: Checklist for identifying the qualities of a network

Methods to support hypothesis formation:
'Social network analysis' is an empirical method that will help you visualise the actors involved in a network and their relationships based on observations (e.g. the number of joint projects and websites) or surveys (e.g. 'Who do you cooperate with?' or 'Who should we contact in this regard in this country?') You can use network analysis and/or other tools to quantify the quality criteria outlined above and examine even the most complex of networks. A number of user-friendly software applications (such as Visione.info) will help you map this information. However, choosing suitable input data and correctly interpreting the findings require long-standing experience.

A map of actors is easier to use in this context, and will help you draft a preliminary analysis. Rather than differentiating between primary, secondary and key actors, the map revolves around the issues at stake in the network. You can use the finished product as a basis for assessing the quality criteria and compiling hypotheses.

Step 3: Plan activities

Here, you identify potential activities for establishing or strengthening a network, based on the goal and the hypotheses identified in the previous steps.

Networking always revolves around creating opportunities for forging new relationships or reigniting existing potential. This includes, for example:

- Activities to maximise opportunities for the participants to establish contacts. Popular networking formats include open space conferences in which the participants themselves decide the programme, or world cafés, where they discuss the issue under review in depth among themselves in alternating small groups.

- Targeted elimination of barriers to communication, e.g. through conflict mediation, creating opportunities for (informal) communication (e.g. fireside chats, study visits) and information material (e.g. newsletters, studies).

- Creating incentives for cooperation (e.g. through small project funds, developing joint knowledge products) or bringing together potential cooperation partners.

- Electronic social networking platforms open up a range of additional networking opportunities. Network members can manage contacts through digital profiles, advise each other in flexible, virtual groups of experts or learn from each other through virtual conferences and learning platforms, providing they have access and know how to use the relevant networking technology.

- Institutionalised network secretariats (e.g. cluster management) can be used to safeguard continuity and drive networking. However, electronic platforms and secretariats may work out to be costly and time-consuming – particularly during the start-up phase – and will never fully replace other activities.

When planning activities in this context, make sure they are in keeping with the four basic characteristics of networks by being needs-oriented, practice-oriented and learning-oriented and complying with guidelines for respectful communication. For more information in this context, please refer to the tools communities of practice and learning networks for multipliers and trainers.

Step 4: Implement activities and monitor results

Although it is often difficult to monitor activities for networks that do not use online platforms, network and/or value creation analyses are just one tool you can use to keep an eye. Social-media-based platforms, on the other hand, are easier to track using a range of activity and networking indicators that are often automatically documented. However you choose to monitor results, it is important that you take into account early on – at the planning stage – how you can measure the degree to which your goal (i. e. establishing or strengthening a network) has been achieved (see step 1) and observe the related indicators. It is important that you review your goal again following an initial phase (of up to six months).

Tool 15
Trust-Building

Notes on use

Purpose	This tool allows you to rapidly assess the existing basis of trust within a cooperation system.[20]
When to use it	When the climate of communication in the cooperation system is disturbed or breaks down.
Setting	In a small group with up to twelve participants; selected steps may also be used as part of a survey.
Facilities and materials	Pinboard with pre-prepared evaluation matrix, document handouts.
Notes	A high degree of openness within the group is a must. Evaluation and discussion of the results require openness. We therefore recommend that you use this tool in a 'safe' setting with small groups. The process can be used either by an external third party, or to promote self-reflection among the actors themselves. Tact should be exercised here. Nobody should be forced to talk about building trust or about any trust issues they may have.

Description

'Trust is the glue of life'.[21]

Trust is an elusive quantity, because it cannot be produced on demand. It grows slowly, is invested and allowed to mature, but can sometimes be lost and tacitly withdrawn. Trust is built on the basis of cooperation experiences and mutual assumptions regarding how other actors involved in the interaction process will behave. When assumptions and cooperation experiences largely correspond, trust grows – i. e. actors project predictable behaviour onto other actors. Trust is a valuable social and economic resource in cooperation systems. It promotes the exchange of information and knowledge, simplifies and speeds up cooperation processes, and reduces transaction costs.

Building trust in cooperation systems is a fundamental prerequisite for effective cooperation. As actors are dependent on each other as regards the change objective, hesitant scepticism, mistrust, tensions and conflicts are major obstacles to effective and efficient cooperation.

Although trust offers major benefits, it also entails risks, as trust can be breached. Over many years, gaming theory has sought ways of resolving with this dilemma. Experiments have shown that one of the most effective ways of dealing with trust issues is basic 'tit-for-tat'. Using this strategy, a party enters into a cooperation system on a basis of trust, but does not hesitate to retaliate in an equivalent manner if one of the other parties is uncooperative[22]. 'Blind faith' or basic mistrust (along the lines of 'Trust but verify') are less-than-promising cooperation strategies.

Building trust is a complex process of communication that requires a considerable investment of time and money. It is not so much the explicit interests of the actors involved that play a crucial role in building trust, but rather their mutual perceptions and assumptions. Actors with corresponding interests may also sometimes mistrust each other. Once it has been built, trust can easily be lost and quickly breached. An irretrievable breach of trust is the biggest obstacle to building further trust.

Trust is built in four ways:

- **Personal experience:** Previous positive and negative interaction experiences are used to make assumptions concerning the future behaviour of the other actor. If the observed behaviour appears predictable and free from harmful intent, trust is invested in the other actor, who accrues social capital. Building trust is a dynamic process in which each side makes assumptions. Large segments of the actors' intentions and opportunities for action remain initially hidden from view, in the background. Misunderstandings can create a huge ripple effect; building trust requires acute awareness. When an actor invests time and energy in building trust through communication, openness and the wielding of influence, without the other actor reciprocating, the investing actor will withdraw, now sometimes more mistrustful than before.

- **Reputation:** The observations and experiences of other parties are used to make assumptions concerning the future behaviour of the other actor. Reputation accelerates the process of building trust. Rather than having to rely on your own experiences, the parties involved can learn from the experiences of others. Online shops such as Amazon and ebay use this resource by asking shoppers to rate their personal interactive experiences and share them with others. Involving third parties not only provides information on whether or not a partner can be trusted. It also gives the partner involved a direct incentive to act in a reliable manner and boost his/her reputation.

- **Sense of identification:** Familiarity with rules and core values make it easier for an actor to make swift assumptions concerning the future behaviour of another actor. Personal traits (such as the age, sex, cultural orientation, charisma, or social class) influence the degree to which trust is invested in an actor. Forging an identity with a group, organisation or culture usually requires a certain level of trust. Identifying common ground (such as joint objectives and other shared traits that may not be public knowledge) helps strengthen the development of trust, as does agreeing codes of conduct and establishing a joint understanding of commitment and fairness.

- **Recognised rules/institutions:** Non-partisan third parties can play a key role in building trust by laying down a cooperation framework or acting as arbitrators. In such cases rather than being invested directly in the cooperation partner, trust is placed in tried-and-tested mandatory procedures that constitute this framework and guard against risks. If cooperation fails, a recognised arbitrator either intervenes or it implements a recognised process of clarification. The trust invested in such an institution is based on its neutrality, predictability, transparency of decision-making, fairness and accountability.

Activities to build trust can target three different levels:

- **Level of individuals:** Trust is not necessarily based on a personal connection. In the first instance, it is based on the assumption that the other actor is positively disposed, and at the very least will not act to harm the actor investing the trust. A wide variety of factors determine the level of trust cooperation partners have in each other. These include good communication skills (the ability to listen and understand as well as a clear vision of one's own interests), respect (openness and a fundamental interest in the other party) and transparent and consistent actions, all of which are fundamental ingredients for building trust.

- **Organisational level:** Individuals are usually seconded from their particular organisation to a cooperation system. This means that they represent the interests of their organisation first and foremost rather than their own interests. It is vital that the interests and expectations of the organisations involved be identified and a reconciliation of interests at this level facilitated ('neutrality'). If the interests of an organisation are not recognised, then the behaviour of the individuals seconded from this organisation is unpredictable. If cooperation is to be based on a spirit of trust, then it is important to communicate with the 'right' individuals. The organisations involved will invest trust in these individuals, who will have the ability to recognise areas where there is scope for cooperation and to represent the interests of their organisation.

- **Level of the cooperation system:** When the boundaries of a cooperation system are being drawn, it is possible to define activities that will help develop trust. It can be useful in this context to involve neutral third parties to resolve any (latent) conflict. It is important to strike a balance between transparency and confidentiality when communicating information externally. Trust is based on the transparent and symmetrical exchange of information on objectives, intentions and plans. Therefore, it is vital that a cooperation system's structures and processes facilitate direct communication, transparent rules and a shared understanding of roles and procedures. A joint understanding of the objective of cooperation is another prerequisite for mutual trust.

When new cooperation relationships are established, carefully thought out activities to build trust play a key role within the cooperation system. Opportunities for informal meetings and encounters are just as important here as transparently structured work processes. A joint excursion by representatives of different organisations, where the ice of mistrust between the actors is given an opportunity to thaw, can be just as important a contribution as an agreement concerning the actors' rights, responsibilities and inputs.

How to proceed

Use this instrument very carefully. Actions speak louder than words, and this certainly applies to trust. This means that you may do more harm than good by asking directly if an actor is trustworthy, or why you should or should not trust another actor. This tool is not designed to eliminate taboos (about problematic relationships). Instead, it will help you come up with ideas for strengthening mutual trust. Paradoxically, even generating ideas requires a high degree of trust and discretion.

Therefore, think carefully about which actors should be involved in using this tool. Be very vigilant. Under no circumstances should use of the tool encourage the formation of cliques, with an 'inner circle' of partners meeting in secret to discuss how to deal with 'difficult' partners. In the worst case scenario, this could destroy a partnership.

Step 1: Define the key focus

In a broad-based cooperation system, you will not be able to study all relationships between actors. Therefore, in this first step, you need to decide which cooperation relationship you wish to examine using the tool. Do you want to analyse mutual trust between all partners in the cooperation system? Or do you wish to home in on specific (bilateral) relationships within the cooperation system. Maybe you wish to shine the spotlight on the relationship between the cooperation system and a key external actor. There are a wide range of options.

Step 2: Assess the cooperation 'climate'

The following analysis of a selected relationship between actors focuses on eight aspects. The total, average and deviation of the rated values (1 to 4) can provide pointers for strategic options and for promoting communication. All participants involved in the analysis can award the number of points they consider appropriate for each of the aspects to be rated.

Positive experiences with cooperation in the past					
Only negative cooperation experiences or none at all	1	2	3	4	Significant, positive and beneficial cooperation experiences
Transparency and predictability of intentions and goals					
Intentions and goals are unclear and concealed	1	2	3	4	Intentions and goals are communicated and clear
Communication among the actors					
There are few opportunities for meeting and communication	1	2	3	4	Regular meetings and intensive communication
Observance of agreements and contracts					
Agreements are ignored and are rarely observed	1	2	3	4	Agreements are negotiated openly and are observed
Fair distribution of advantages and gains					
Advantages and gains are acquired unequally	1	2	3	4	Distribution is openly negotiated and a fair solution is found

Trust in the representatives of the other actor					
The behaviour of representatives is arbitrary and changeable	1	2	3	4	Representatives know each other and work to maintain good relations
Conflict management					
Tensions and conflicts are not talked about or addressed	1	2	3	4	Conflicts are addressed openly and constructively early on
Public image of the relationship					
The image is one-sided and disadvantageous for us	1	2	3	4	The agreed image strengthens our relationship and is positive

Working aid 18: Assessment of the cooperation climate

Step 3: Compile hypotheses

In this step, you compile hypotheses on the following questions based on the assessment of the existing cooperation climate.

- What are the strengths of the relationship on which trust can be built?
- What (latent) conflicts might make trust-building more difficult? Could there have been a breach of trust?
- What influence does the personal relationship between actors have on trust or lack thereof?
- What influence do different organisational interests exert?
- What influence do existing agreements (or lack thereof) exert on the cooperation system?

Step 4: Assess the features of successful partnerships (optional)

Using working aid 19 may help you gain further insight when compiling hypotheses. It will assist in pinpointing latent conflict and identifying any gaps in the design of the cooperation system. It is important to recognise such deficits, as otherwise they will give rise to conflict that could be attributed solely to the individuals involved rendering trust-building activities utterly pointless.

Features of successful partnerships	Assessment		
	absolutely	partially	not at all
Individuality All cooperation partners contribute something that is of value to the others, but remain autonomous.			
Significance of cooperation The cooperation relationship is important to the participating actors (the individuals involved and their organisations).			
Interdependence The cooperation partners complement and need each other; none can achieve alone what all can achieve together.			
Investment The participating partners mobilise the resources available to them, and in so doing demonstrate their interest in partnership.			
Communication The cooperating partners keep each other informed and make use of opportunities for exchange. Tensions and conflicts are addressed early on.			
Integration The cooperating partners offset imbalances of information and participation.			
Learning Periodic evaluation of experiences and joint success stories are made visible.			
Institutionalisation The cooperation relationship is cemented through a minimum of agreed, useful rules.			
Integrity The cooperating partners behave with integrity, openly keep each other informed, and in so doing deepen mutual trust.			

Working aid 19: Features of successful partnerships

Step 5: Plan trust-building activities

Here, you plan trust-building activities based on the hypotheses you have compiled. These activities are designed to strengthen the cooperation climate and ensure success by creating an atmosphere in which working with other actors is enjoyable.

The key question in this context is 'What conditions do you need to put in place to ensure that partners cooperate with each other in a spirit of trust'?

Raising the relevant actors' awareness and encouraging them to make an active effort will help significantly improve the situation. If necessary, you may need to consider that a smouldering conflict may be undermining the formation of trust.

The following working aid will help you identify and plan activities in different fields.

Field	Planned activities
Level of individuals, e.g. ■ More opportunities for one-to-one (informal) discussion ■ Mediation by external individual (third party) ■ Coaching/training for support ■ Reallocation of responsibilities	
Level of organisations, e.g. ■ Clearer objectives ■ Better knowledge of existing interests and expectations ■ Greater reconciliation of interests ■ Involvement of other/additional representatives from one (or more) organisations	
Level of cooperation systems, e.g. ■ Establishment of a joint coordination platform ■ Definition of binding standards for processes ■ Identification of joint milestones ■ Formulation of rules for conflict management ■ Definition of joint activities for mutual trust-building ■ Evaluation of experience at periodic intervals; visualisation of joint achievements	

Working aid 20: Activities for developing partnerships

Step 6: Implement activities and reflect on results

It is vital that you also jointly implement any activities that you have discussed with other partners or that, at the very least, you identify and come to a binding agreement on who will do what. It is also important that you agree with those involved that any information gleaned in the context of using this tool be treated confidentially. You should also reach a consensus on how the results of the activities implemented will be monitored and on what reflection process will be used.

Tool 16

Backstage and Learning Behaviour

Notes on use

Purpose	This tool will help you obtain a deeper understanding of patterns of action within a cooperation system (unspoken rules, problem-solving behaviour, learning behaviour), and allows you to draw conclusions on possible focal areas to strengthen the system.[23]
When to use it	Generates deeper insights into the cooperation system, can help you understand and address patterns of behaviour in situations where no explanation can be found at the superficial level.
Setting	Workshop.
Facilities and materials	Spacious room (commensurate with the size of the group), circle of chairs or groups of tables (coffee shop setting), space for several working groups, pin-board, flip chart, workshop materials, document handouts.
Notes	Before you address this theme in a workshop setting, participants must be willing to engage in critical reflection on their own patterns of behaviour. Make sure you allow sufficient time to ensure this is the case. The steps described below are just one possible sequence. Ask yourself this question before you start: What do you wish to achieve with the outcomes/insights attained? Alternatively, you can use the tool to document the cooperation system's patterns of action.

Description

Cooperation as a theatre production

We can gain a new perspective by looking at a cooperation system involving numerous actors as if it were a theatre production. On the stage we see the actors play their roles, represent their interests and develop their relationships with other actors. They build relationships of trust, negotiate joint ventures, place the carefully built trust at risk, and plunge unexpectedly into deep conflicts. They act according to their roles, scripts, expectations, influence and resources, and with the other actors and their relationships in mind. They build a network of interdependencies. There are shifts in the power and influence wielded by the actors. Many things lack transparency and are kept concealed, because actors wish to be seen in a certain way. Information is exchanged and withheld; rumours are spread. Actors are prompted to remember their lines, scenery is shifted, and backstage the casting of actors in their various roles is negotiated. Actors put on their costumes, props are distributed, and scripts are rewritten. Deceit and intrigue come into play, and strings are pulled from above. Some actors move into the spotlight, others remain in the shadows.

The otherwise quite limitless creative possibilities are limited and steered by the actors them-selves. They are the ones who create the structural conditions on stage which enable them to develop their relationships. Like theatre productions, cooperation systems create an inner world of implicit rules of which the actors are only partially aware, even though they themselves have created them. The actors' problem-solving behaviour and capacity to learn and manage change play a crucial role in shaping and influencing this inner world and the performance capability of the cooperation system.

Understanding cooperation systems as theatre productions calls for detachment, self-critical re-flection and a broad vision, because perception is not a purely technical and rational operation. Perception is selective, and involves components of projection, active organisation, attribution of meaning and construction of form.

Choosing our perception

Our attention focuses on things that confirm what already fits into our world view, do not gen-erate any dissonance with our beliefs and cultural orientations, and appear useful to us. Led by our utilitarian outlook, we tend to believe that people will learn something in order to be able to use it later on. This is not necessarily true in all cases. People can enjoy a moment of learning as a welcome opportunity to encounter something new, and then return to the order of the day when the new experience no longer has a role to play.

Projection

We constantly project our desires and meanings into what we perceive. We assume that others share our perceptions, and are astonished when they do not. This becomes especially apparent in the interpretation of symptoms. Symptoms are observable and perceptible signals from which the status and dynamics of the cooperation system can be inferred. Symptoms need not neces-sarily point to a deficit or a weakness. They may also reflect strengths and potentials. The diverse impressions that we gain from a cooperation system compel us to read patterns and trends into incomplete data. Symptoms are always tied to the people, processes, structures or products of the cooperation system. Individual symptoms in themselves tell us little or nothing. The symptom 'lack of orientation and weak leadership' in a cooperation system may have unintended posi-tive effects on the development of exploratory initiative among the actors. In another case the same symptom might speed up the disintegration of positive elements, or lead to a situation in which useful potentials go to waste. We must therefore always check the symptoms and analytical findings together with the actors of the cooperation system. This process of reflection allows us to assess the differing significance of symptoms, because the same symptom can be interpreted differently. We can interpret precision as pedantry, managerial overview as a compulsion to con-trol, creativity as chaos, discipline as rigidity, reliability as perfectionism, unreliability as cheerful flexibility, and discretion as wilful suppression of the truth.

Organisation and construction of meaning and form

The contents of perception are placed in relation to each other. This operation is shaped basically through the distinctions offered by a language. Perception and expectations, interpretations and intentions are harmonised to form a stable lifeworld which enables us to communicate about our actions. These viable fundamental beliefs form a reservoir of tacitly shared assumptions that combine to form a social framework in which the sense or meaning of things is constituted. New experiences are added to the framework. People have the ability to read a pattern into incomplete, fragmentary information, and to construct an overall picture from this. The ability to capture this kind of structure is based on a process of associative thinking that is rooted in the lifeworld, and obeys an economic logic: We don't need to know everything in order to make sense of various items of information and data, and be able to act accordingly.

How to proceed

To approach the various realities within a cooperation system it is useful to apply various comparative perspectives. We will confine ourselves here to three perspectives:

- the issue of the implicit rules backstage of the cooperation system

- the problem-solving behaviour within the cooperation system

- learning behaviour within the cooperation system

Step 1: Analyse unspoken rules

There are no hard and fast guidelines for identifying the unspoken rules. The route that takes us backstage leads across front stage, i.e. through everything that is tangible and visible in relation to priority setting, preferences, relationships and influence. To approach the micropolitical inner world of a cooperation system it is helpful to imagine the difference between a long-standing and a new member of that system. New members are not familiar with the unspoken rules. These rules are not announced at introductory meetings, nor are they written down in any handbooks or guidelines. New members of an organisation or a network need plenty of time to reach an understanding of how it really works.

The following key questions, raised in one-on-one or in working group situations, help identify the unspoken rules:

Preferences

- What are the three most visible and measurable aspects of our cooperation system? – Examples include efficient communication, high transaction costs or visible products.

- What needs or themes does the cooperation system need to address more intensively?

Relationships

- What was decided on within the cooperation system, but was either not implemented at all or only implemented very slowly?

- Who considers the coordination mechanisms within the cooperation system to be competent, and who is dissatisfied with them?

Backstage – the subtext

- What would you advise a new member of the cooperation system to avoid doing and saying?

- What would you advise a new member of the cooperation system to do and say if he or she wanted to be accepted and to fit in as quickly and as smoothly as possible?

Step 2: Assess problem-solving behaviour

The problem-solving behaviour of cooperation systems has proved a key factor in practice. Any analysis that takes this fact into account will therefore seek to answer the following three key questions:

- What practical experiences of change do the actors in the cooperation system possess? – Examples include internal restructuring and process acceleration, development of relationships with other organisations and cooperation management, development of new products and services.

- What problem-solving methods do the actors have at their disposal? – Examples include problem-solving groups, knowledge management, peer-to-peer learning (intervision), workshops.

- What attitude do actors adopt to approaching problem-solving and the performance of new tasks? – Examples include routine, curiosity, openness, reserve.

Step 3: Analyse learning behaviour

As different actors interact, cooperation systems create a fundamentally conducive learning climate through the exchange of information and knowledge, and horizontal cooperation between the actors. The learning orientation can, however, be constrained by veto power.

This step is therefore designed to identify the possible wielding of veto power, so that targeted measures can be taken to resolve the situation. The items outlined in the following table can assist you in this context. They can be worked on either by individuals or jointly within the group. It is also possible to make a note of the everyday observations on which a rating is based.

Veto 1: 'Silo' mentality					
Little contact among the actors, high degree of specialisation among small groups that communicate in codes.	1	2	3	4	Actors work in separate groups simultaneously, are horizontally networked and work on joint projects.
Everyday observation:					

Veto 2: Confusing architecture for innovation					
No clearly communicated innovative themes, dissipation in numerous uncoordinated initiatives and workshops with isolated channels of communication.	1	2	3	4	Clear priority setting for innovation in a few thematic areas, bundling of initiatives, mainstreaming has high priority, all pull in the same direction.
Everyday observation:					

Veto 3: Authoritarian and ideological blinkers					
Fixed articles of faith are repeated mantra-like, deviating opinions are frowned upon, critique is risky.	1	2	3	4	Actors are invited to contradict, critique is called for and rewarded, and actors are willing to experiment.
Everyday observation:					

Veto 4: Time pressure, heavy workload					
High density of rules wastes time, actors have heavy workload and are under time pressure.	1	2	3	4	Actors perform routine tasks with ease, scope is created for maintaining relationships and for new tasks.
Everyday observation:					

Veto 5: Communication gaps					
Actors are not well informed and use information gaps as a power resource, there are few opportunities for exchange.	1	2	3	4	Actors are well-informed and communicate proactively.
Everyday observation:					

Veto 6: Unutilised experience					
Evaluation of experience is an onerous duty and a special task, nobody is interested in results.	1	2	3	4	Evaluation of experience is an integral part of the work process, actors evaluate experience periodically and utilise it.
Everyday observation:					

Working aid 21: Analysing learning behaviour

Step 4: Analyse and draw conclusions

Analysing the three perspectives – unspoken rules, problem-solving behaviour, and learning behaviour – will help you gain a deeper understanding of the cooperation system. In a next step, you can draw conclusions about possible development paths and potential approaches to promote and strengthen:

- individual actors, e.g. through integration and participation, improved conflict management skills, enhanced performance capability for project or financial management, or for cooperation with other actors in the cooperation system;

- individual relationships between actors, e.g. by promoting information and knowledge exchange, confidence-building forms of encounter, communities of practice, utilisation of information and communication technologies, incentives for new forms of cooperation;

- the cooperation system as a whole, e.g. by supporting and mediating processes to negotiate norms, rules, cooperation guidelines, enhanced performance capacity of coordinating cores and improved evaluation of experience.

Tool 17
Needs Analysis

Notes on use

Purpose	This tool allows you to establish where you require complementary cooperation and to identify the internal or external partners with whom you should forge such a relationship.[24]
When to use it	When you require additional cooperation inputs.
Setting	Workshop with key actors.
Facilities and materials	Flip chart, pinboard, workshop materials.
Notes	You will require a good knowledge of the project and of the permanent cooperation system.

Description

Exploring possible complementary partners usually opens up a project's system boundaries. Here, the focus is on which add-ons and partnerships would be particularly desirable for the project.

Before you can define your exact needs, you must have a clear picture of the project's performance profile. You can only identify what is lacking and where a complementary partnership would be of further assistance if you assess the project's (limited) performance capacity realistically, against the backdrop of the agreed objectives and results.

To explore your need for external support, you need to define various criteria and draw up a specific performance profile for the project. Visualising this profile highlights possible gaps and provides specific pointers as to what you need to look for in external complementary partners.

How to proceed

Step 1: Assess the project

Five examples of criteria that you could use to assess the project are provided below. You can, of course, modify the wording as you see fit or use other criteria. Start by formulating an assertion for each of the criteria and then rating it on a scale of 0 to 3:

- 0 = absolutely false
- 1 = partially true
- 2 = mostly true
- 3 = fully true

Financial resources: The project has sufficient financial resources to achieve the targeted objectives and especially to scale up the solutions in other contexts and use them in political dialogue. – Rating e. g. 2.2

Knowledge and expertise: The project has access to current knowledge to achieve the targeted objectives. – Rating e. g. 1.8

Access to opinion-makers and political decision-makers: The project has good access to the relevant opinion-makers and political decision-makers. – Rating e. g. 1.6

Scaling-up and/or regional dissemination: The project feeds the solutions developed into the political dialogue effectively, and the solutions are also applied in other contexts. – Rating e. g. 2.1

Implementation competence: The project has sufficient implementation competence. – Rating e. g. 2.2

Step 2: Visualise the findings

Enter the described values along the lines that represent the respective criteria, and join up the points. This will produce a performance profile for the project (ACTUAL profile).

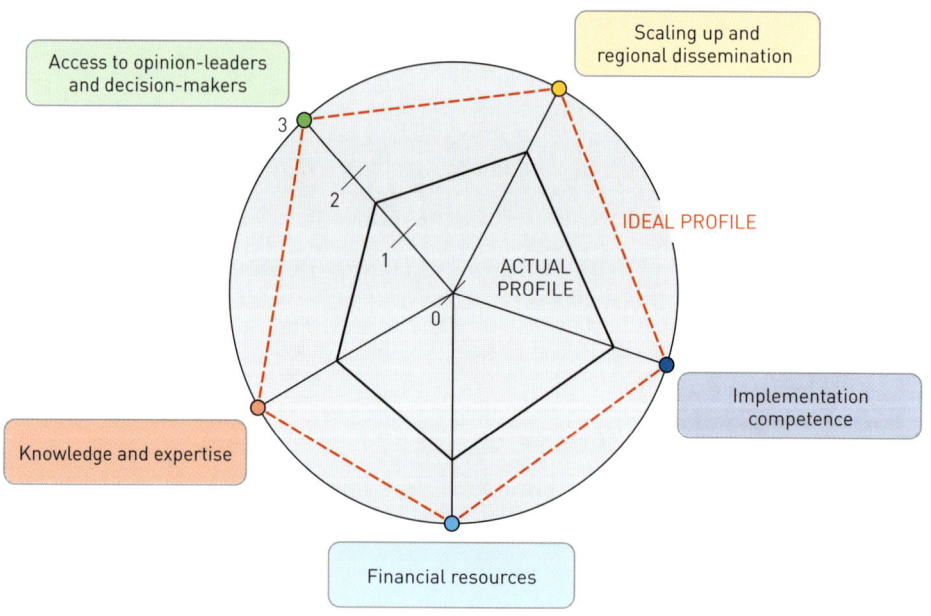

Figure 37: Project performance profile

Step 3: Interpret the findings

Figure 37 shows the difference between a project's IDEAL profile and its ACTUAL profile. Discussion of the findings shown in the example for instance would lead to the following conclusions:

- You need additional knowledge and expertise.

- You need support in applying specific solutions (regional scaling-up).

- You lack access to opinion-leaders and political decision-makers.

- You do not require additional financial support.

- The project is strong on implementation and does not need any external partners.

Follow-on question: Which potential complementary partners could you use to fulfil the identified needs? For example:

- You could enter into a partnership with various public and private research institutions to access additional knowledge and expertise. You could also outsource this knowledge management task.

- You could set up a committee comprising representatives of various institutions which could be used as a political network to improve access to relevant opinion-leaders and political decision-makers and boost the leverage of the project.

Tool 18

Comparative Advantages

Notes on use

Purpose	You can use this tool to help structure and describe the comparative advantages of a project that make it an attractive partner for other actors.[25]
When to use it	In situations where a project needs to motivate additional external or internal partners; in marketing situations.
Setting	Workshop with key actors.
Facilities and materials	Circle of chairs, possibly space for working in small groups, pinboards, flip charts and workshop materials
Notes	Demands a user-oriented perspective on the project and internal expertise. This tool is not designed to identify comparative advantages; it does, however offer a structure for describing them.

Description

A potential cooperation partner will only be attracted to a project if it believes that it will receive something in return that is proportionate to its input to the cooperation system.

To ensure that this is the case, the project must have a clear picture of what its comparative advantages are and how it is going to make them visible. It is of no benefit if the project itself knows that it can offer special expertise in certain areas if others do not.

How to proceed

Step 1: Discus comparative advantages

This tool is designed to stimulate reflection on your own core competences within the context of a project. This reflection should be more than just a standard description of what a project intends to do. To obtain a profile of a project's core competences, you need to discuss three key questions in relation to the following areas:

A. Strategies and concepts

- What strategies and concepts set you apart from others?
- How are these strategies and concepts linked to overarching discussions in the field of development cooperation?
- How do these strategies and concepts help make cooperation more effective?

B. Management and steering

- What tools and methods for steering projects and cooperation are you able to offer?

- What are the distinguishing features of these tools and methods?

- Where do you possess sound expertise and in-process experience in different contexts?

C. Evaluating experiences and learning

- What participatory method for monitoring and evaluating experience do you have at your disposal?

- How can you design evaluation and learning processes that allow you to feed back authentic experience into the project's strategic orientation?

- How do you support learning and change processes in organisations?

Step 2: Outline the comparative advantages

On the basis of the responses to these key questions, draw up a short one-page outline of the project's core competences. The key question here is: What do you have to offer? You should answer this question as specifically as possible, and back up the response with examples, where possible.

Here is an example of an outline for a decentralisation project.

Strategies and concepts

We help structure and deliver advisory services for political dialogue on the decentralisation agenda. In the negotiation and harmonisation of interests between interdependent public and private actors and between national reform programmes and donors, we pay close attention to three points:

- access to new knowledge on practical models of decentralised administration;

- support for consultation with actors at various levels, especially on fiscal issues;

- involvement of bodies for institutional political participation: parliament, commissions, associations, political parties and trade unions.

We attach special importance to clearly defining decision-making levels and strengthening results-based development project management and steering capacities at the central, provincial and municipal levels. We also promote donor harmonisation in specific areas: public financial management including public revenue and the court of audit.

Management and steering

We have introduced participatory, negotiation-based decision-making mechanisms among partners and donors in order to be able to jointly steer the project. We therefore promote the participation of governmental and non-governmental actors in political processes in particular, in order to strengthen ownership at all levels. We possess sound expertise and in-process experience in the:

- introduction of a proven instrument for participatory municipal budgeting;

- strengthening of cooperation and network management within and between municipalities;

- delivery of advisory services on internal management issues within the administration at all levels (human resource development, financial management, etc.);

- strengthening and acceleration of internal administration processes;

- strengthening of sectoral competences and leadership.

Evaluating experiences and learning

We attach particular importance to leveraging local experience in municipalities, as this helps feed authentic information into the political negotiation process. We support partners in establishing a project portfolio management and monitoring system.

We promote supervisory and monitoring mechanisms to improve accountability and social responsibility.

Step 3: Prepare presentation of marketable core competences

Preparing a brief outline of core competences provides external partners with an overview of why they should enter into a partnership with the cooperation system. You will be able to refer to it when contacting possible external partners.

Tool 19
Shaping Negotiation Processes

Notes on use

Purpose	This tool will help you prepare and implement negotiations in which different interests, ways of working and expectations meet and perhaps clash.
When to use it	Wherever bilateral or multilateral agreements need to be reached/decisions need to be taken.
Setting	Between two and twenty participants.
Facilities and materials	Circle of chairs, possibly space for working in small groups, pinboards, flip charts and workshop materials.
Notes	Using this tool requires a high degree of openness and diplomacy. It cannot be used mechanically. The art of negotiation is usually only acquired through corresponding training and experience.[26]

Description

Negotiation processes in cooperation systems

In cooperation systems, decisions are always taken in processes of negotiation between the cooperation partners. Unlike organisations, cooperation systems have no line management level that can make a decision on behalf of cooperation partners if something is unclear or there are differences of opinion.

When cooperation partners wish to pursue a shared objective, they become interdependent. This applies to all forms of informal and contractually formalised partnerships. At the same time, the individual participants also pursue their respective individual objectives, and have their own methods for performing their work, to which the other participants have only limited access. Successful negotiation is therefore a key prerequisite of successful cooperation.

Negotiation processes are always happening in a cooperation system. Therefore, you cannot measure the success of an individual negotiation situation solely on the extent to which a negotiation partner was able to assert itself. In this context, successful negotiation must always fulfil additional functions, namely strengthening joint objectives orientation, achieving a better understanding of the individual interests involved and consolidating the overall cooperation relationship. Even if you only negotiate once with a cooperation partner, any accusation of having acted unfairly in this context could tarnish your reputation as a cooperation partner in the long term. So, good negotiation skills involve more than 'simply' asserting yourself.

The shaping of interdependent cooperation relationships depends basically on how the participants negotiate, how they exchange knowledge and how they achieve solutions that generate
benefits for all the participants, or at least the majority of them. Imbalances in the distribution of
benefits lead to increased expectations among those who walk away from the negotiation emp-
ty-handed, or who feel disadvantaged. This means that a partnership will gain stability over a
prolonged period of time if and when the negotiations lead to a balanced distribution of benefits.
It would be wrong to think that any scenario can or must be transformed into a win-win situation
through negotiation. As the saying goes, 'You win some, you lose some'. The most important thing
is to acknowledge this fact and ask how it will be dealt with in the long term.

In negotiations between cooperating partners, clashes occur between different interests – interests
that are legitimate from the point of view of the parties involved. It is therefore important to establish a culture of communication where the cooperation partners not only want, but are also able,
to maintain and continue good relations with each other after the negotiation. You can achieve
this by conducting negotiations so that hardened negotiating positions can soften, interests can be
openly addressed and new potentials for creative agreement are harnessed. The starting point for
this is acknowledging the legitimacy of different concerns and interests. Then, if the concerns and
interests of all participants are identified, there is a chance that a number of new agreements can
be reached. To explore these possibilities, you must allow the negotiation process to constantly
discover and assimilate new aspects.

In a nutshell, negotiation involves:

- acknowledging the different interests of the parties involved

- broadening the system boundaries during the negotiation process to include new elements

- precisely defining various interests and the advantages and drawbacks of different solutions

- developing solutions that are better for the parties involved than no solution at all or leaving
  the cooperation partnership

Basic principles of negotiation

Although this logic may be compelling, real negotiations never follow this pattern exactly. Negotiations are influenced by the time and place, but above all by the participants themselves, who
influence the process with their various cultural orientations, capacities and more or less transparent strategies. How the negotiation proceeds depends on how successfully you can structure the
process together with the participants. Within any negotiation process, the structure of the process itself is also always an object of negotiation. The less structured the process, the more urgent
the need to recall the following basic principles for negotiating partnerships:

Objectivity

Cooperation partners usually tend to attribute the problem (such as scarce resources, poorly defined responsibilities or conflicting objectives) either to one actor's position or to the relationships between the parties involved.

The drawback of this is that the negotiation then tends to address individuals and their positions rather than the actual issues, and the objective of the negotiation shifts towards changing the behaviour of the other cooperating partner. This, however, runs counter to the basic idea of negotiation between cooperation partners who are different and who pursue different interests on a legitimate basis. Successful negotiation in cooperation systems must therefore seek to focus attention on the objective issue at stake, e.g. scarce resources or poorly defined responsibilities. When pursuing that path, it will also be necessary to make detours. It may prove necessary for the participating partners first of all to express their mutual perceptions, and even vent their feelings of annoyance.

Identify and acknowledge different interests

The participating partners in cooperation often tend not to formulate their different interests clearly. Cultural orientations and the need to create a semblance of harmony can lead to actors behaving as if all participants had the same interests. Recognition of the fact that different cooperation partners have different interests, and the right to assert those interests in negotiations, has first of all to be achieved in the course of the negotiation process. The parties involved usually figure out in advance whether they have an alternative to negotiation, and at what point and under what circumstances they might want to leave the negotiation process. The possibility of making concessions is not fixed, but may be deferred based on the acknowledgement of interests, or the joint identification of new aspects in the negotiation process.

Broaden the scope of alternative options

The more precisely you can shed light on the object of a negotiation from various sides, the more new information flows into the process. It is helpful here to listen to and take on board the expert opinions of third parties. This broadens the options available, stops actors from fixating on what they see as the only conceivable solution, and creates new perspectives. Ideally, this will mean identifying a range of possible solutions that the negotiating partners will at least recognise as such.

Agree on assessment criteria

To evaluate these different solutions, the negotiation can focus on agreeing assessment criteria. These should shed light on the possible benefits for the cooperating partners, and the long-term consequences of the solution. Finally, on the basis of these transparent solutions it then becomes possible to consider issues of compensation in order to reconcile divergent interests.

How to proceed

Based on the principles outlined above, the following **six negotiation steps** have proved useful in practice.

Step 1: Deal with the different positions and interests

To generate an ethos of cooperation based on fair play, it helps if you start by separating positions from facts. In this first step, all parties declare their respective interests in the issues being negotiated. For a negotiation process to be successful, all of the parties involved must make their own positions, as well as their reasons for holding them, clear and transparent.

Step 2: Deal with the issue

In this second step, you can analyse in detail the issue to be negotiated together with the other parties involved. During this step, it may well be appropriate to obtain further information, or listen to the opinions of experts. The feeling of having obtained new information together with your partners paves the way for the next step.

Step 3: Explore mutual interests

Looking at mutual interests will help you secure sound negotiation outcomes. What do the parties involved wish to achieve together? What visions of the future do they share? In this step, you pave the way for creating mutual trust.

Step 4: Develop alternative options

In the next step, you should avoid reaching premature solutions. Instead, utilise all the new information available on the issue at stake and develop alternative options. Fostering this process of creative thinking is worthwhile. As new options emerge both sides will become more confident that all they need now are appropriate criteria for selecting one of several alternatives.

Step 5: Agree on assessment criteria

In this step, you agree on criteria for assessing possible solutions. You can round off the joint evaluation and selection of possible options by skilfully introducing compensatory elements wherever there is any remaining sense of unease.

Step 6: Select an option

Once you have assessed the possible options, the one that is most suitable for the parties involved is chosen. You then agree on additional implementation activities.

The following checklist will help you work through the six steps outlined above.

Issue to be negotiated	Action item
Separate positions from facts: declare interests, explain reasons for your own position.	
Obtain new information jointly on the issues being negotiated: analyse subject matter in detail, listen to experts.	
Identify shared interests, build trust, clarify mutual expectations.	
Develop alternative options: use new information on the issue at stake.	
Agree on assessment criteria: evaluate solutions and consider compensation.	
Select solution, agree on what action is to be taken.	

Working aid 22: Checklist for the negotiation process

Success factor Steering structure

Tool 20

Steering Structure

Notes on use

Purpose	This tool will help you develop, select and decide on a suitable steering structure for a project.
When to use it	To ensure transparency and clarity about responsibilities and roles, and to form the basis for ownership. Ideally, you should decide to adopt a particular steering structure at the beginning of a project, or when a project is being strategically reoriented. Establishing a steering structure is a major intervention in the project and into the permanent cooperation system. The point in time at which you establish the structure, and the depth of intervention, should therefore match the project's present stage. When a project is launched, for example, simple steering structures may be adequate, whereas once it has evolved further and there is greater need for decision-making, you may need to establish deeper structures.
Setting	Workshop with key actors.
Facilities and materials	Pinboard and flip chart, workshop materials (markers, cards, etc.), if available: map of actors, document handouts.
Notes	You will require a good knowledge of the actors in the area of social concern to help ensure that the right people are involved. This tool usually requires preparatory interventions to help ensure that the cooperation partners are comfortable with their roles.

Description

Any project is a temporary cooperation system. Each project therefore needs its own, tailor-made steering structure to supply it with decisions. There are no blueprints, because projects operate within different organic structures and coordination mechanisms, which should be taken into account when elaborating the steering structure.

When projects design steering structures they are able to draw on tried-and-tested models that have gained acceptance within the respective organisational cultures. These can then be adapted to the specific needs of the project. The steering structure must often fit into very diverse organisational landscapes (e.g. in projects that are jointly designed by actors from the public and private sectors and from civil society). At the same time, you should try to select a steering structure that is conducive to innovation. Since steering forms the inner core of a project, the steering structure selected has enormous ramifications. It sets the framework boundaries for what can be learned in the project and how: communication patterns, interaction schemes, participatory processes, consultation and decision-making processes, knowledge management and the design of work processes for learning.

Experience has shown that we should distinguish between politico-normative, strategic and operational levels of steering. This distinction relieves high-ranking decision-makers, for instance, of having to take decisions that can be taken by people at the next level down who have better access to the relevant information. Applying the subsidiarity principle in this way also ensures greater overall acceptance of the steering structure among the actors involved.

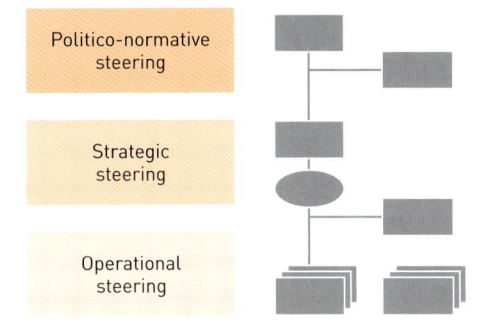

Figure 38: Steering levels

How to proceed

Step 1: Identify possible participants in the steering structure

In this first step, you need to identify possible participants in the steering structure by conducting an analysis of actors. From the point of view of a cooperation system, it is desirable to have as many steering topics as possible dealt with on a participatory basis involving the relevant actors. This is not always possible in practice.

The focus here is on participants who:

- make political decisions;

- are responsible for achieving objectives and sub-objectives;

- provide an important impetus for achieving sub-objectives.

A good way to start is to involve the key actors who are identified in the course of an analysis of actors in the steering structure (see the success factor cooperation). The discussion can be facilitated by visualising the structure. The size of the circles in the diagram below can be used to represent the participants' presumed degree of influence on steering decisions.

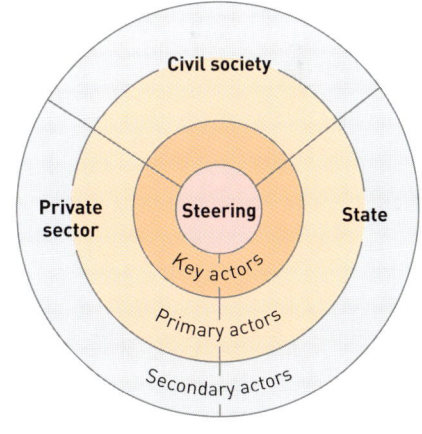

Working aid 23: Identify possible participants in steering

Step 2: Identify steering tasks

Before defining appropriate forms of participation in the steering structure it will help if you consider carefully the functions of the steering structure and translate these into steering tasks such as strategy, planning, coordination, control, monitoring, resource management and conflict management. These tasks will need to be planned for by allocating sufficient resources and assigning responsibility.

Different forms of participation may be called for, depending on whether the steering tasks are allocated to the politico-normative, strategic or operational level. Therefore, it is important that you define and describe the steering tasks involved.

Step 3: Determine the forms of participation in the steering structure

In this step, you develop different forms of participation and levels of intensity, based on the complexity of the task in hand within the project. If the intended form of participation is not developed and communicated in an appropriate manner, participants in steering will usually think of themselves as being on a higher level of intensity than was intended by the cooperation partners inviting them to participate. In other words, instead of simply taking note of a decision they will question it; instead of accepting an explanation they will offer good advice on what solution would be better, and so on.

We can distinguish between the following forms of participation in the steering structure:

- **information** on decisions provided through the normal channels (e. g. newsletters, websites, project reports);
- **information and communication:** communication of detailed information on decisions and explanation of the reasons why they were taken (e. g. project presentations, events);
- **consultation** prior to decision-making or participation in preparatory work for decision-making (e. g. focus groups or sounding boards);
- **participation** in the decision-making process (e. g. working groups);
- formal **direct responsibility** for decisions (e. g. steering meetings).

For each steering task identified in step 2, different forms of participation can be assigned to the actors involved.

In the matrix, it is a good idea if you enter the frequency of participation and the time input required (e. g. regular weekly meetings, two steering group meetings per year, one-off survey). This will highlight the opportunity costs of steering: steering consumes time and energy.

Steering task:					
Possible steering participants	**Participation level**				
	Information	Information +	Consultation	Participation	Responsibility
Actor 1					
Actor 2					
...					
Actor n					

Working aid 24: Participation levels

Step 4: Define the politico-normative level

The politico-normative level is the level at which the objectives and the fundamental values and rules of conduct within the cooperation arrangement are negotiated and laid down. The achievement of objectives is monitored here and any adjustments to the objectives are agreed as required. Fundamental conflicts of interest or violation of shared values are dealt with at this level.

All projects are embedded in an area of social concern. Within that field, a decision-making body will usually already have been set up at the politico-normative level. The project should link up with these existing structures, in order to sustainably tie the temporary cooperation system into the area of social concern.

Should no such body exist, the participants should agree to set up a body of their own. Where such bodies do exist, but function ineffectively, the actors involved should agree on improving these bodies. Projects should generally be cautious about setting up bodies of their own, because any new body will weaken existing steering structures.

Step 5: Define the strategic level

The strategic level determines which path the project will take in order to achieve the objectives. At this level, the steering structure maintains an overview of progress and deviations from targets during implementation, reflects on strategic options and agrees on milestones for further implementation. In other words, this is the level at which management challenges within the project are dealt with.

The process of agreeing on the composition of the steering structure for the strategic level should not be geared exclusively toward participation by the key actors. This is the level at which all the elements of a project come together. It often plays a pivotal role in ensuring that political directives, strategic decisions and operational implementation fit together. The members of a corresponding body should therefore possess managerial expertise – particularly of a strategic nature – and have sufficient time to be able to supply the project with the decisions required for achieving the objectives and results.

Step 6: Define the operational level

The operational level assumes responsibility for all the day-to-day decisions needed to implement activities within the prescribed strategic framework. It provides a decision-making basis for the strategic level, by supplying information on progress and on deviations from targets during implementation.

The design of the operational level will depend on how a project is to be implemented. The following options are conceivable, and hybrid forms may also be selected:

- **Thematic structuring**
  The operational level is structured along the lines of the various themes that have a bearing on the project. (For instance, in a project to promote sustainable economic development, the operational level could be structured along the lines of microfinance, value chains, vocational training, etc. as areas of intervention).

- **Regional structuring**
  The operational level is structured according to the regions in which the project operates to ensure integrated steering of the various interventions (for instance, one operational unit for province A, one for province B, etc.).

- **Structuring by phases**
  The operational level is structured so that it meets the requirements imposed by the various phases of the project. (For instance, if the project design provides for a pilot phase followed by scaling-up, then it would make sense to design the operational level for the pilot phase differently than for the scaling-up phase).

- **Structuring according to learning needs**
  If the rapid results approach is employed, the temporal structure will be a different one. Smaller-scale and shorter learning projects will be launched that enable lessons learned at the operational level to be used to design a wider programmatic approach at the strategic level. The operational level will then be structured to match the rhythm of the learning projects.

Step 7: Describe roles, responsibilities and processes

Once the basic elements of the steering structure have been agreed, you need to describe in detail the roles and responsibilities of the individual bodies at the various levels. In complex steering structures, it is extremely important for all partners working in the cooperation system to know and understand who is involved in decision-making processes at which levels, what their mandates are, and their role in decision-making. We therefore recommend laying down in writing, for example, for each steering level how often the relevant bodies will meet, what decisions they will take, who the members of each body will be, their mandates and roles, and how the interfaces between the steering levels will be designed.

Tool 21
Qualities of a Steering Structure

Notes on use

Purpose	This tool will help you reflect on, develop and optimise a suitable steering structure. It lists nine requirements that you can use to assess the functionality and appropriateness of an existing or planned steering structure.
When to use it	To plan and reflect on an existing project. It will help you to visualise communication and decision-making structures in the project and to address the quality requirements of the actors involved.
Setting	Workshop with key actors.
Facilities and materials	Pinboard and flip chart, workshop materials (markers, cards, etc.).
Notes	General use of this tool requires an understanding of and experience in project steering. You will also need to ensure that there is sufficient openness within the group.

Description

All projects need a steering structure to provide a framework for communication and decision-making processes. There are no blueprints, because projects operate within different organic structures and coordination mechanisms. It is important that you ensure that steering structures for temporary cooperation systems are geared to existing structures or decision-making bodies of the permanent cooperation system to the greatest extent possible in order to prevent the creation of parallel structures. Ultimately, you will only be able to judge the quality of a steering structure using two criteria: an optimal structure must be functional with respect to the targeted objectives and results, and must be appropriate for the complexity, variety and the scope of the task in hand.

At a practical level, the actors involved have at least an implicit idea of how steering should be structured for each project. However, actors' ideas on the structures, rules and roles that are to be communicated and decided upon in a project often diverge radically, even in projects that are already underway.

This tool provides you with nine requirements that will allow you to visualise and systematically discuss actors' ideas.

How to proceed

Step 1: Outline the steering structure

In this step, participants are asked to visualise the existing or planned steering structure for all to see. If you anticipate there to be radical differences in interests in the structure or in how those involved perceive the existing structure, have an initial draft compiled by small groups or by individuals. In a second step, get the participants to compare and combine the different versions.

- Agree on whether participants are to outline the actual situation (how does the structure look now?) or the target situation (how will the structure look in future?). If there are any doubts, start by depicting the existing structure.

- Outline the project's joint communication and decision-making structures. In other words, focus on communication between the cooperation partners, not on communication in individual organisations.

- Depending on the initial issues at stake, decide whether participants are also to depict informal parts of the structure. Start by drawing up a list of the 'social spaces' that are key communication platforms for the project (i.e. formal bodies, informal lunch meetings, etc.). Early on during the preparatory stage, weigh up the potential for greater scope (offered by involving steering structures that are taboo or that were not previously considered) against the risks that could ensue by breaking taboos.

- You can use different visualisation forms such as:
  - organisation chart: depicts formal bodies such as steering groups, sounding boards and working groups as boxes arranged hierarchically. This is the format usually chosen as it is easy to understand. Drawbacks, however, include the lack of a temporal dimension (and a dynamic structure) and of informal elements.
  - visualisation of 'places' of communication and decision-making over the course of time (see the intervention architecture tool)
  - matrix: visualisation of roles within the project (state actors, representatives of NGOs, private sector actors, etc.) and their tasks in sub-projects and structural elements.
  - flow of communication: visualisation of the communication flow between structural elements and/or roles in the project.

Step 2: Reflect on the steering structure

The nine requirements listed in the following working aid will help you reflect in detail on the steering structure. It is impossible to completely fulfil all of the nine requirements at once. For example, a wide variety of perspectives in steering may lead to greater conflict sensitivity, but it will also lower efficiency. A very high level of transparency in steering could also have a negative impact on proactive behaviour.

If necessary, you can define a preliminary structure for discussing the nine requirements by choosing only those requirements with the greatest leverage for improving the steering structure.

Requirement	The steering structure ...	++	+	–	––
Proactivity	... strengthens ownership and self-reliance of actors at the different working levels. This is achieved primarily through participation and negotiation orientation.				
Transparency	... strengthens confidence in steering through broad-based communication on decisions and the transparent criteria on which these were based.				
Efficiency	... is simple and makes it possible to reach decisions without excessive transaction costs being incurred for consultations, negotiations and coordination.				
Variety of perspectives	... takes into account the different perceptions and perspectives of actors. It combines both hard data and reports on individual experiences as well as different interpretations of these data.				
Conflict sensitivity	... makes it possible to identify tensions and conflict at an early stage and to work these through. This is ensured in particular by applying the do-no-harm principle.				
Flexibility	... allows a quick response to changes in the environment, strategic reorientation and financial restrictions, for example.				
Mainstreaming	... uses existing structures and coordination processes in the area of social concern.				
Learning from patterns of action	... allows actors to take on new roles. It helps them practise new patterns of management action, for instance in communication and decision-making.				
Organisational model	... serves as a model, as it generates innovative stimuli in the organisations involved. Elements of the steering structure are also used outside of the project.				

Working aid 25: Checklist of nine requirements

Step 3: Draw conclusions

In this last step, you draft and agree on activities and options for further developing the steering structure based on discussions. Documenting the outcome may help you to subsequently inform actors who were not involved in the process.

Tool 22
Results-Based Monitoring System

Notes on use

Purpose	This tool provides you with an overview of the steps required to set up a results-based monitoring system.
When to use it	A results-based monitoring (RBM) system should be available as soon as possible for project steering. You can use it to set up an RBM system at the start of a project and to guide you through operations.
Setting	For set-up: two to three-day workshop with an internal working group that includes key actors who have experience in setting up and operating monitoring and evaluation systems. Using an external moderator with experience in monitoring and evaluation will prove useful.
Facilities and materials	Flip chart, pinboard, workshop materials.
Notes	A sound knowledge of the project, its objectives, the context and general conditions is key. The results model should be available. Ideally, you should also have the information provided by a map of actors and a process map for the project, and the project's steering structure should be operational.

Description

Results-based monitoring (RBM) is one of the key steering tasks in a project. The RBM system will help you continuously review progress in achieving jointly agreed objectives and results and to take corrective action where necessary. It is an essential component of project operations and implementation. Without RBM, steering is like flying blind.

All projects need to implement RBM so that at any time they:

- … can access information on the project's progress (verification of results);

- … know what works and where changes are required (learning);

- … make strategic decisions based on monitoring data (steering);

- … initiate dialogue on the chosen strategy and the plan of operations with the actors involved (communication);

- … have a reliable basis for fulfilling accountability obligations (reporting, evaluation).

The tool comprises six process steps that describe how the RBM system is structured and used. It provides practice-oriented, methodological guidance on operationalising an RBM system.

Ideally, a joint monitoring planning workshop will be held with the relevant actors to elaborate process steps 1 to 4. During the workshop, plans of operations are drafted for monitoring activities that are then documented in a monitoring instrument.

Results-based monitoring and Capacity WORKS

Capacity WORKS and RBM are closely interlinked. Monitoring provides information on the results that activities achieve, thereby allowing you to draw conclusions on the strengths and weaknesses of the activities implemented. Weighing up the success factors will help you gain a better understanding of why some activities achieve the desired results while others do not.

The success factors will also help you formulate indicators that provide information on the quality of cooperation and show whether the project is on the 'right' path to change.

- Strategy: Do the actors involved have a common understanding of the project strategy as a path to change? Can the cooperation partners share ownership of the joint objectives and desired results?
  Possible indicator: Commitment and financial engagement by key actors

- Cooperation: Which actors must be involved in the project in order to achieve the objectives and results? Who will take on what roles and responsibilities?
  Possible indicator: Routine review of the map of actors in a steering group meeting

- Steering structure: Which actors are key to project steering to ensure that the jointly agreed objectives and results are ultimately achieved? How are decisions taken?
  Possible indicator: Quality of steering decisions in case of conflict as perceived by all actors, or achievement of milestones from the plan of operations

- Processes: Which core processes in the area of social concern need to be the focus of attention to ensure maximum leverage? How can the processes in the joint project (core processes, steering processes, support processes) be coordinated as closely as possible to ensure that the envisaged positive changes in the sector are achieved?
  Possible indicator: Quality of internal output processes

- Learning & innovation: Who has to learn what, and at what level, in order to achieve the objectives and results and to ensure that the required development capacities are mainstreamed in the area of social concern in the long term?
  Possible indicator: Implementation of routine learning events in the project, e.g. four times a year.

How to proceed

Step 1: Devise, review and adjust the results model

In this first step, you draw up a results model for the project if one does not already exist and if one does, you review and revise it (cf. the results model tool).

Step 2: Clarify the requirements of the RBM system

In this step, you clarify the requirements of the monitoring system. This involves mainstreaming the monitoring system in the project's steering structure so that it can provide the relevant actors with the information required for making decisions that will drive the project's progress.

The following questions will help you clarify the requirements:

- Which actors are to be involved in the key strategy and steering decisions to be made by the project?

- How are key steering decisions made and what information is required to this end?

- What interests, expectations and information requirements do the different actors have with respect to a joint monitoring system?

- What information must the monitoring system be able to provide, and when?

- Which actors are to be involved in monitoring? Who is responsible for which aspects of monitoring?

- Do the cooperation system partners possibly already have monitoring systems in place that can be used as a basis for (improving) the joint project?

- What human and financial resources are required for setting up and operating the monitoring system? What resources are available?

Step 3: Make results measurable

Here, you make the results defined in the results model measurable. To do this, you review the underlying hypotheses and adjust and supplement them where necessary. In this step, you also need to define indicators in order to measure whether the project's planned objectives and results are being achieved.

Indicators are a crucial element of any monitoring system. The efficiency of a monitoring system depends first and foremost on the quality of the indicators defined. Indicators are reference values that give specific information on complex issues and allow them to be measured. They show whether and to what extent a planned quantitative or qualitative change has occurred. Bear in mind the following quality criteria when formulating indicators:

- They must be objectively verifiable (i.e. they must be SMART – specific, measurable, achievable, relevant and time-bound).

- Indicators are results-oriented. In other words, they should describe what results will be achieved, not how they will be achieved.

- They must have a verifiable baseline and a target value (benchmarking).

- Indicators must contain as much clear detail as possible about the data sources or data collection methods and this should be used when determining indicators (verification).

Step 4: Draw up detailed monitoring plan and set up the monitoring instrument

In this step, you draft a detailed monitoring plan for the entire project term and channel the outcomes of steps 1 to 3 into a monitoring instrument (e. g. an Excel or web-based tool).

The monitoring plan should contain all of the required processes, steps, methods (e. g. for data collection), deadlines (e. g. data collection schedule/measurement intervals) and responsibilities for ongoing monitoring.

The monitoring instrument should structure the data collection processes and the systematic documentation of the collected data. It should also make it easier to interpret and use the data to steer the project.

Step 5: Collect and (routinely) analyse the data

Here, you routinely collect and evaluate the data.

Collect the following information for all of the indicators and enter it in the monitoring instrument:

- baseline data/target value/milestones;

- actual values (at the agreed time intervals);

- an assessment of the degree to which the objectives and indicators have been achieved.

Step 6: Use the findings of RBM

The aim of this final step is to use the findings of RBM:

- for ongoing steering (strategic, managerial and budget-related decisions etc.) and for mainstreaming RBM in the decision-making mechanisms of the steering structure and of the actors involved;

- for accountability, substantiation of results and evaluation obligations and for reporting;

- for in-project knowledge management, documentation and communication and for supporting sustainable learning processes.

Tool 23
Architecture of Intervention

Notes on use

Purpose	Preparing an architecture of intervention will help you to plan and steer interventions. It indicates along the time axis which actors will meet in what form in order to work through the joint issues.[27]
When to use it	To draft transparent plans for implementing agreed interventions. It will also help you to document planned processes so that they can be discussed with others, reviewed at regular intervals and adjusted.
Setting	Suitable for working alone or ideally also in pairs with another individual involved in the intervention who has experience of the process, or with an external consultant; can also be applied in a small group if tightly moderated.
Facilities and materials	Note pad; possibly flip chart or pinboard with cards.
Notes	You will need as clear a picture as possible of the requirements of the project, as well as a precise knowledge of the involved actors and frameworks. Ideally you will have prepared a map of actors beforehand. This tool is not suitable for planning an entire project. It is, however, useful at the level of individual, manageable sub-projects or lines of action.

Description

Whereas the plan of operations provides a detailed description of who will carry out what activities and when, the architecture of intervention provides a structured, one-page overview of the objective, content-related, social, temporal, spatial and symbolic dimensions of the planned interventions, thereby placing all the information in a single coherent context.

This overview should correspond with milestone planning.

Objective and content-related dimension

In terms of the project strategy (capacity development strategy, results model), many decisions are taken implicitly, such as those concerning the anticipated duration of the project, modes of participation and so on. The issue of who is to contribute objective knowledge to the project is also crucial in this context.

The social dimension

This dimension is the most significant of all for the architecture itself. It determines who is to be involved in which interventions, and in what form. Participation revolves around the agreements within the framework of the success factors cooperation (forms of cooperation) and steering structure (levels of the steering structure).

The degree to which political actors and their divergent interests need to be integrated will partially determine the number of meetings, the number of different social spaces and their size. Energy for change is generated along the margins of systems. In the architecture of intervention, the management ensures that people and organisations sit together and cooperate in new configurations.

Possible design elements within the social dimension include:

Symbol	Nature of intervention	Functions
	Steering group meetings	decisions, strategy, planning, coordination, control, monitoring, resource management, conflict management, reflection, catalyst (for new ideas, initiatives), addressee (for questions, information)
	Working group meetings at the level of operations	design of concepts, preparatory reports for decision-making and implementation, prioritisation, information platform, broadening of participation
	Workshops (e. g. kick-off, diagnosis/monitoring/ evaluation workshops)	evaluation, results-based monitoring, management of relationships of trust, creation of a basis for joint decision-making on further procedure (e.g. various working groups/participants in implementation)
	Advice, coaching of managers	alleviation of pressure within the system through stabilisation of individuals (managers have a role model function), raising of credibility through involvement and strengthening of the management level
	Dialogue groups	coordination between key actors, mutual information and communication requirements, eliminating mistrust, misunderstandings and undesirable developments
	Dialogue meetings with other actors (large-scale events)	creation of a group feeling, creation of purpose, inclusion of key knowledge bearers and perspectives through dialogue, fresh stimulus for the system
	Training	development of the required capacities, generation of commitment among participants, depending on composition of group: promotion of sharing/team building, stabilisation of the system through capacity building

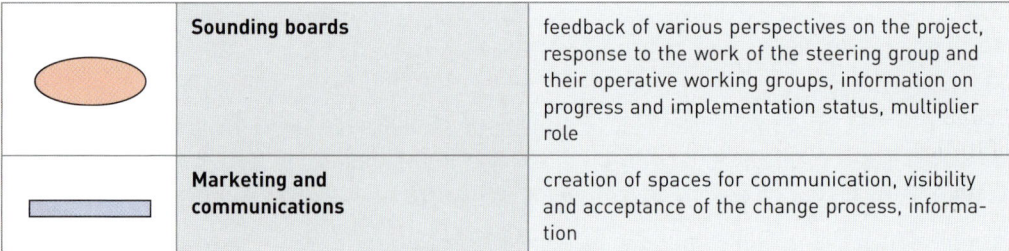

	Sounding boards	feedback of various perspectives on the project, response to the work of the steering group and their operative working groups, information on progress and implementation status, multiplier role
	Marketing and communications	creation of spaces for communication, visibility and acceptance of the change process, information

Figure 39: Possible elements for designing an architecture of intervention

The temporal dimension

The work and decision-making phases are arraned in a timeline to fit the tasks in hand, and the social elements are then added to each phase. The timeline also allows processes to be speeded up and slowed down. There is one rule of thumb for interventions: key things must happen within six months (concept in place, financing secured, specific issues dealt with, etc.), otherwise motivation and people's ability to remember (energy levels) will subside. The timeline of the architecture of intervention is aligned with milestone planning.

The spatial dimension

The spatial dimension is one that is usually neglected, even though it can have a very significant impact on the context at an implicit level. The term 'spatial' in this context refers to the location of events (within or outside the locality of the working system), to seating arrangements and to whether or not these arrangements can be designed spontaneously (festive or sober design of spaces).

The symbolic dimension

Symbols activate mental maps that provide meaning and help orient action. The use of symbols will supplement your linguistic repertoire, which is why it is also referred to as symbolic language. People usually also notice whether the verbal and symbolic language being used is consistent, or whether conflicting messages are being sent. Where the two do conflict, people tend to give more credence to the symbolic language, because it is often (correctly) assumed that this represents the unconscious component. It is therefore especially important to harmonise the two. It is very important at the beginning or at the end of projects to consciously manage the symbolic language used. It is also an unsurpassable medium for transitions from one phase to another or for expressing appreciation. The introduction of rituals – which are used to pass on organisational knowledge – also constitutes part and parcel of setting the scene for symbolic language.

Examples include:

- Meaning: at the beginning of an important process a representative of the top steering level attends a half-day meeting in order to underline the importance of the process. Before leaving, he/she provides brief feedback on what he/she will be taking away from the meeting.

- The completion of a sub-task is emphasised symbolically (topping-out ceremony, etc.).

How to proceed

Step 1: Define the design context

Before you start to design an architecture of intervention, you should first of all remind yourself of the main contextual aspects:

■ What are the objectives? What expectations are there as regards the time frame?

■ Who is participating in the process with what interests and/or who will be affected by the results?

■ At which locations can work take place with which relevant individuals?

■ Which actors must meet? When? How?

■ Which tasks must be completed? Which roles must be performed within the process in order to achieve the objective?

■ What hypotheses exist concerning the process design and implementation?

Only when a clear picture has been obtained of the framework within which an objective is being pursued does it make sense to think about designing the process. If there are still issues to be clarified, it is advisable not to begin with a detailed architecture of intervention straight away.

Step 2: Define core elements of the architecture

Based on this brief analysis of the context, you should be able to design a rough framework for the architecture of intervention. It is helpful here if you think about which key actors need to meet in order to discuss and work on the key topics. It is also often helpful to enter roughly along the time axis certain milestones or dates that are of special significance (interim outcomes, events, deadlines for negotiation and decision-making). This gives you a rough diachronic framework with windows of time that you can now fill in appropriately.

When drawing up an architecture of intervention, it is also a good idea if you start off with a rough outline before moving on to compiling a more detailed architecture. Remember too during this step that you need to take into account corresponding steering elements and elements to support the project, and not just the elements for designing the actual content of the process.

To document the architecture of intervention, also for others, it will help if you visualise it by placing the symbols for the various process elements along a time axis (see Figure 40 below). You can also place the different types of process elements on separate lines, making the figure easier to read.

Here is an example of an architecture of intervention:

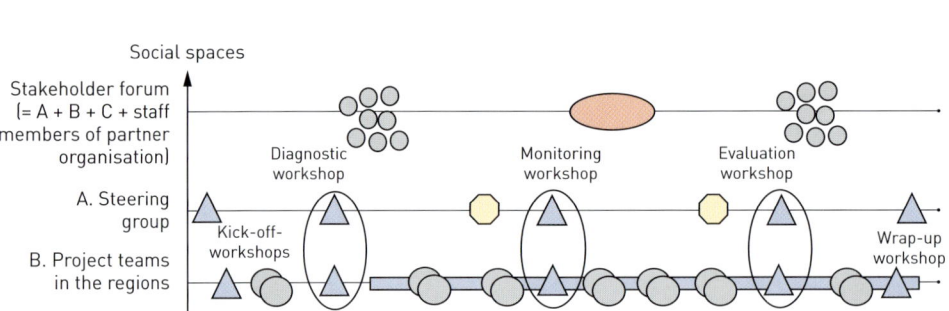

Figure 40: Example of an architecture of intervention

You can also create new design elements, or adapt the ones proposed here to your own context. The symbols each indicate which intervention is planned for which social space at which point in time. So, to roughly define the architecture of intervention you need to answer the following key questions:

▪ What is the composition of the core team/key working group? At what intervals are they required to meet for sequences of work? What key events need to take place?

▪ How will the process be steered? Which actors will be involved in the steering structure, take key decisions and provide the process with the necessary legitimacy and resources?

Step 3: Add detail to the architecture

Only when you have put in place a sound rough architecture is it appropriate to start detailed planning. We therefore recommend that you incorporate corresponding supplementary measures for instance before and after the defined key elements of the process. Prior to events involving large groups it might for instance be appropriate for you to incorporate corresponding information and communication processes. After events, you should create scope for evaluation and the formulation of conclusions. At key points for decision-making, you may need to listen to a sounding board. Labour-intensive phases or steps may require appropriate support or the inclusion of a broader resource base. In this way, you can add appropriate design elements to the architecture of intervention until you feel you have designed a sufficiently sound process.

To review whether the process is sufficiently sound, it may be useful to answer the following key questions:

- Does the design contain a sufficient number of work sequences?
- Are the planned steering elements adequate?
- Are those affected integrated early on and in a positive manner?
- Does the architecture contain the elements needed to support the work itself?
- Are there elements that perform a quality assurance function?

Step 4: Review and adjust the architecture continuously

A further step is not only to simply implement the architecture of intervention once it is in place, but also to review and if appropriate update it at relevant points in time. The context often changes during the course of a project. New actors come into play, restrictions are lifted or the strategy changes in response to changes in the project setting. With this in mind, we recommend that you review the architecture of intervention at regular intervals, and adjust it as appropriate.

Tool 24
Plan of Operations

Notes on use

Purpose	This tool will help you to agree on specific implementation arrangements with the actors involved once the project strategy has been defined.
When to use it	In situations where you need to develop and document specific actions and activities for implementing projects.
Setting	A workshop with key actors and the actors involved in implementing activities.
Facilities and materials	Pinboards, workshop materials (markers, cards, etc.).
Notes	Before you use this tool, you need to have a clear understanding of the project's strategic orientation, drawn up together with the relevant actors who bear joint ownership.

Description

A plan of operations is a document that identifies key packages of tasks, decisions, responsibilities and milestones for implementing a strategy over a specific time frame. A time frame of one year is generally advisable. It sets out who will do what and when.

Operational planning is a management task and includes fundamental decisions on the output processes within the project. Planning operations means designing and planning these output processes, i. e. channelling scarce resources into efficient procedures, outputs and work packages.

In the project's steering structure, the strategic orientation, input commitments and complementarity of the project's outputs are harmonised. Final decisions on this are made during operational planning.

When planning project operations, you must bear in mind that all of the partners involved have at the back of their minds the (planning) logic they apply in their home organisations. You must assume that there will be conflicts of objectives between the organisations represented and the project. Therefore, planning operations in the project presents both a challenge and an opportunity to:

- translate the strategic **priorities** into outputs and work packages;
- promote cooperation between the actors through a joint **approach**;
- establish **transparency and balance** between project-based and organisation-based planning among all of the cooperation partners involved;

- achieve successful, binding and trust-based decision-making on the **allocation of resources**;

- generate **synergies** with the cooperation partners' strategies of action.

How to proceed

Operational planning is an iterative and recursive procedure. In the past, a workshop format involving all the relevant project partners has proven effective.

You should most certainly involve individuals who are involved in the project at the strategic level. It is also helpful to involve people from the operational level who will be responsible for implementing the agreed work packages. Select the participants very carefully as the more people are involved, the more complex the process becomes.

You can apply the following steps to a project as a whole, or – if that is too unwieldy – to segments of it (e.g. lines of action).

Step 1: Take stock of the preceding period

If available, you should discuss and analyse the findings of the periodic monitoring and evaluation activities. What has been achieved? What still needs to be done?

Experience shows that it is worthwhile linking up this first step of operational planning with the results-based monitoring process (RBM workshop).

Step 2: Check the strategy

You now need to review the project strategy, and if necessary develop it further. The following questions will prove helpful:

- What do you wish to achieve?
  Here, you should refer to the objectives, the targeted results (e.g. in the results model) and the capacity development strategy. Review the strategic orientation and check that it is up-to-date, write down the strategic priorities for the period being planned (if appropriate for individual lines of action) and consider the risks.

- How can you achieve it?
  This question points to the strategic themes in the success factors cooperation, processes, learning & innovation and steering structure. The perspectives of the various success factors will help you integrate the managerial aspects of project implementation into operational planning.

At the end of this step, draw conclusions for planning, based on the review of the preceding period and the strategy check. What are the strategic objectives for the current planning period (e.g. one year)? What indicators will help you establish that these objectives have been achieved?

You can enter objectives and work packages for the period to be planned in working aid 26. This will give you a rough overview of what tasks need to be carried out.

Strategic planning	
Strategic objectives with indicators	**Work packages**

Working aid 26: Strategic planning: Objectives and work packages

Step 3: Plan milestones and activities

In this step, you flesh out the work packages identified in step 2 and channel them into activities that will be implemented as part of the corresponding work packages during the next planning period. Here, you:

- plan activities;

- agree on milestones (point in time by which the activity should be completed);

- appoint the persons responsible;

- roughly assign resources and budget.

Plan of milestones				
Work packages	**Activities**	**Milestone (point in time)**	**Responsibility**	**Resources and budget**

Working aid 27: Plan of milestones

Step 4: Work out plan of operations and allocate resources

In some cases, the planned milestones will provide a sufficient basis for implementation. Those responsible for implementation will then take care of more detailed planning in their specific area of work.

In other cases, it is helpful if you add a more detailed plan of operations. To do this, you work out in precise detail the specific activities required to achieve the planned milestones. You can use working aid 28 to do this.

Step 5: Document and feed the outcomes into the RBM system

In the preceding steps, you documented the entire plan of operations. This will provide you with an important basis for implementation and for monitoring the project's effectiveness:

- review of the preceding period;
- strategic plan (including a strategy check);
- plan of milestones;
- plan of operations.

Remember to document the milestones in particular in the RBM system and to monitor them regularly.

Plan of operations – Planning period:

Project/line of action/work package
Result:

			Schedule												Respon-sible	Use of resources		Cost of materials	Other costs	Com-ments
#	Activity	Indicator/ interim results	J	F	M	A	M	J	J	A	S	O	N	D		Personnel – from the project – other personnel				
																Project	Addition-al per-sonnel			
⋮																				
⋮																				

Working aid 28: Plan of operations

Success factor Processes

Tool 25

Process Map

Notes on use

Purpose	This tool will help you visualise the relevant processes within a cooperation system.
When to use it	To help you develop a strategy by mapping the permanent cooperation system based on an analysis of the current situation. The portrayal of the temporary cooperation system will also help you steer the project and prepare for operational planning.
Setting	Workshop with relevant actors.
Facilities and materials	Pinboards, workshop materials (markers, cards, etc.).
Notes	You will need a sound knowledge of the permanent cooperation system in order to be able to identify and assess the processes. A strategy must be in place before you map the project's processes. The objectives of the project must also have been clearly identified. The process map will help you take stock and to organise processes in cooperation systems. The tool is not intended for designing processes in the strict sense. Although you use it primarily to depict processes within the project, you can also use it as a basis for fine-tuning process design.

Description

When you use this tool, start by deciding on a focus: What do you want to depict? The permanent cooperation system (area of social concern)? Or the temporary cooperation system (the project)?

Mapping the processes of a permanent cooperation system allows you to analyse the current status of the area of social concern and to:

- identify or clarify the 'raison d'être' of the area of social concern (as regards the provision of services for a society);

- assess the achievement of objectives in the permanent cooperation system (access, costs, quality);

- identify responsibilities and mandates in the area of social concern;

- draw up an overview and take stock of processes in the permanent cooperation system;

- identify the need for change in the area of social concern;

- devise possible points of entry for a project.

Mapping the processes of a temporary cooperation system allows you to:

▪ identify responsibilities and mandates in the project;

▪ take stock of the existing processes and the interfaces that exist between the different processes within the project;

▪ draft indicators that are geared to the quality of cooperation within the project;

▪ identify the project's optimisation potential.

The process map is a proven tool for identifying and visualising the steering, core and support processes in relation to the objectives of the area of social concern or the project.

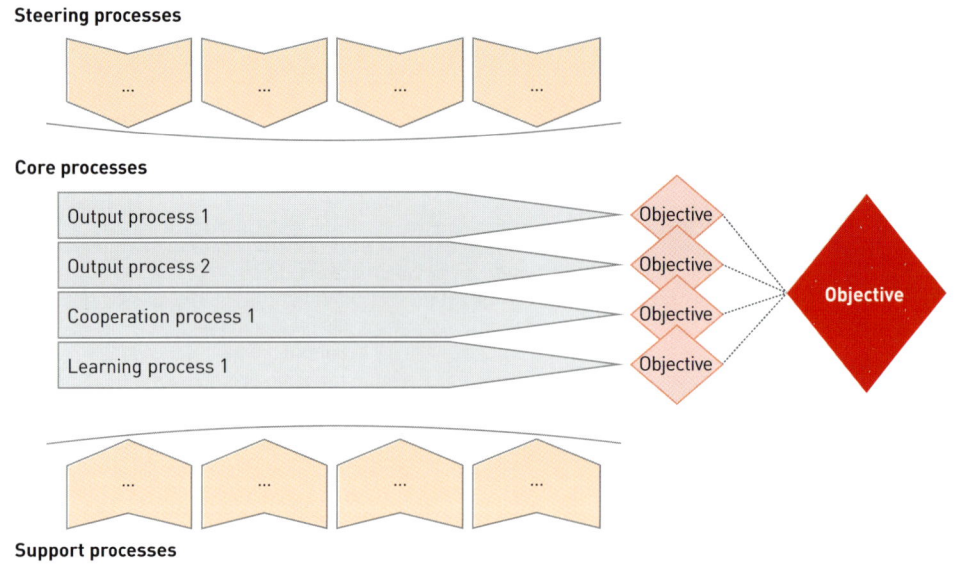

Working aid 29: Process map

Directly or indirectly, all processes in the permanent cooperation system help ensure that users benefit from an output (objective), for example by opening up access for the population to health facilities. If the process map relates to a project, the processes identified contribute directly or indirectly to achieving the agreed change objective.

We recommend breaking down processes into different types both within the permanent cooperation system and within the project:

Output processes
are processes that generate added value, i. e. a visible benefit, for the recipient of an output.

Cooperation processes

include all processes that allow various actors within the cooperation system (e. g. bodies, working groups) to cooperate, and to consult and coordinate activities. Well-designed and properly structured cooperation processes allow actors to continue performing their own particular tasks while at the same time coordinating their activities. They also facilitate the efficient use of existing resources and allow the actors involved to avoid unnecessary duplication, and to overcome barriers to communication.

Learning processes

enable actors to appraise the quality of output delivery and make necessary changes. Well-designed learning processes allow participants within the cooperation system to learn from experience and to continuously improve their performance. Individual learning is every bit as important as organisational learning and learning within the cooperation system.

Core processes

Output, cooperation and learning processes are closely intertwined and have a direct effect on the quality of output delivery within the cooperation system. This is why they are referred to collectively as core processes.

Support processes

are packages of tasks that underpin all the other types of processes. They have no direct contact with output delivery and often, one of their characteristic features is that they can also be farmed out to external service providers. This maintains the long-term availability of the output processes (e. g. knowledge management, implementation of training measures, etc.).

Steering processes

set the legal, political and strategic framework for the other process types. They provide orientation through targets, resource allocation and regulations (e. g. laws, strategies, public budgets, etc.). Steering processes supply the cooperation system with decisions.

How to proceed

Step 1: Define objectives

You derive the objectives from the objectives system of the area of social concern or from the project's targeted objectives and results. Regardless of the perspective to be viewed, you start by agreeing on objectives, either from the point of view of the permanent cooperation system or from the point of view of the project.

Situations can arise in which no clearly formulated objectives for the permanent cooperation system can be identified, or in which several objectives exist that conflict with each other. This information is highly relevant when seeking to understand an area of social concern and provides pointers as to why specific processes are proving ineffective, for example. You should therefore also document this information.

Mapping a project's processes involves reaching agreement on joint objectives as part of the nego-tiation process between the actors involved in the temporary cooperation system. For this reason, you should wait until a strategy has been developed before mapping the project's processes.

Step 2: Identify core processes

The procedures involved in the following steps are the same for both types of process map (for permanent and for temporary cooperation systems). In each case, the focus is on the objectives, which guide your perspective. Asking the following questions will help you identify and structure the processes:

- Do the objectives of the cooperation system already imply or suggest that certain core pro-cesses already exist?

- Which output process will help you achieve objectives?

- Which processes ensure that the actors involved in the cooperation system cooperate with each other and coordinate their activities?

- Which processes support joint learning and the generation of innovations? How are experi-ences exchanged? How is quality assured?

In order to identify the core processes, it may help if you compile a list of all sub-processes before assigning them to the corresponding process types.

Step 3: Identify support processes

Answering the following questions may help you identify the support processes:

- What are the key support processes needed for core processes to run smoothly?

- Which support processes help deliver the outputs required within the cooperation system?

- Which support processes help the cooperation system to function?

- Which support processes make simple, high-quality learning possible within the cooperation system?

Step 4: Identify steering processes

To identify the steering processes, you first need to determine what strategic and legal frameworks already exist for the core processes (in the form of targets, resource allocation and regulations). Then you need to ascertain which steering processes will help further differentiate the frameworks that provide orientation and facilitate decision-making.

Step 5: Prepare the process map

Once you have identified all of the core, auxiliary and steering processes, you can incorporate them into the process map. This will provide you with a clear overview that will help you draft hypotheses regarding how things stand in the area of social concern or in the project.

When you create a process map, the exact assignment of processes to individual process types is not as important as the discussions to agree on a joint map of relevant processes. Start out by simply depicting the objectives and processes. There is no need to assess them just yet. Once the processes have been mapped, determine where processes are lacking, are not harmonised or are not yet working efficiently. You can then move on to identifying what changes are required.

Tool 26
Process Hierarchy

Notes on use

Purpose	This tool breaks down key processes into their sub-processes, specifying their content. This will help you describe the processes in the process map in greater detail.
When to use it	In situations where you require a more detailed process description to establish: What individual steps do you need to carry out? Which actors will be involved? Which interfaces do you need to take into account? What is the objective of the process? The process hierarchy will come in useful when using the outcome of strategic discussions among high-ranking cooperation partners at the operational level, e. g. to draft a plan of operations for a project.
Setting	Workshop with participants in the process.
Facilities and materials	Pinboards, flip charts, markers, PowerPoint and video projectors.
Notes	You do not need to describe all processes in detail in this way. Start with the issue to be addressed. Focus on the processes that are of key strategic importance. You should already have compiled a process map. You will require a detailed knowledge of the relevant processes, along with access to information about costs. The implementing actors rather than the decision-makers need to be involved here.

Description

This tool will help you to break down processes that have been analysed at a strategic level into specific process steps. For example, for processes that were developed within the framework of strategic plans, the tool will help project managers to break down these processes into their components and individual processes, step-by-step. You can break down the processes to the level of operational planning. This will help you establish a logical connection between the strategic analysis and detailed operational planning.

Process hierarchies may also prove useful when mapping processes in an area of social concern. You can use them to analyse selected process steps, which will help create a better understanding of how things stand in the area of social concern, where necessary.

How to proceed

Step 1: Clarify the issue at stake

You should compile a process hierarchy for a specific issue. What information do you need a process hierarchy to deliver? On this basis, determine which process steps will be analysed in greater detail.

Step 2: Define the process hierarchies

Start this step by selecting one or several key process steps for analysis. Break down each of these steps into a process in its own right, and into individual steps. Continue until you reach the desired degree of detail for the specific issue.

Avoid going into too comprehensive a degree of detail. The complexity of the processes involved can quickly overcomplicate things. It is better if you focus on the processes that are the most important strategically or key to the issue in hand.

The following working aid illustrates this principle:

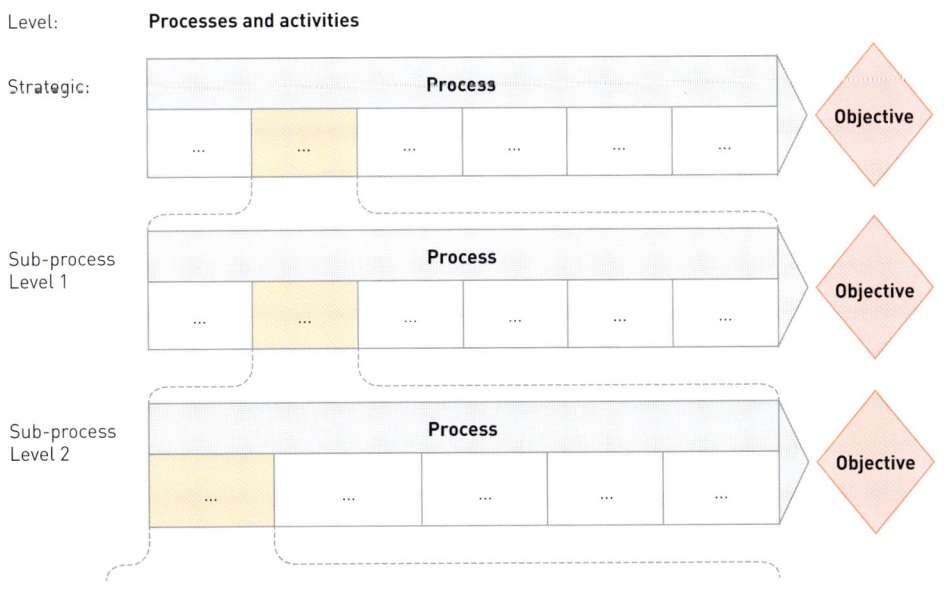

Working aid 30: Process hierarchy

Here is an example taken from a watershed management project in the Mekong river region:

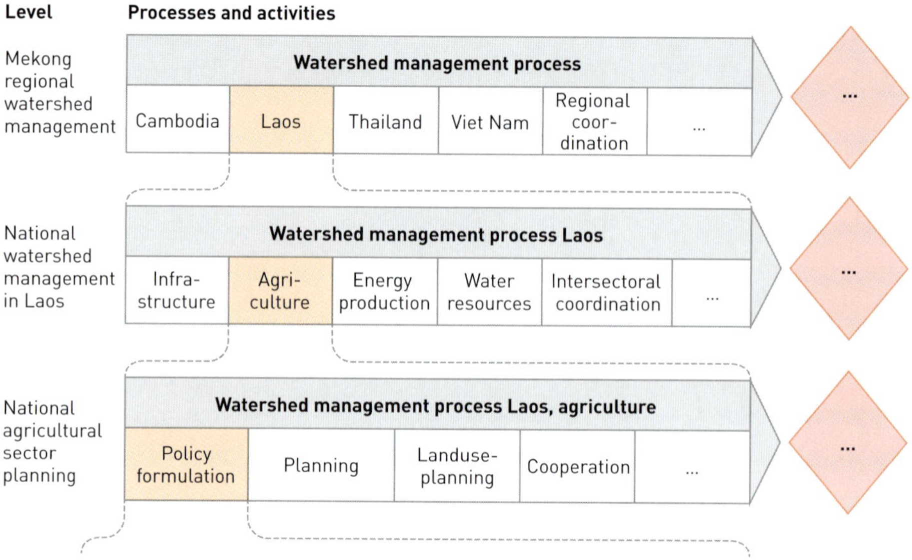

Level 1: To ensure that a regional watershed management system is a success, the watershed management pro-
cesses in the participating countries must interact to a sufficient degree.
Level 2: You can describe the structure of these watershed management processes for individual countries.
In this example, the situation in Laos is described.
Level 3: In this case, agriculture is a key pillar. You can describe key processes.

Figure 41: Example of process hierarchy for watershed management

Step 3: Draw conclusions

When compiling a process hierarchy, we recommend that you focus on the processes that are the most important strategically or key to the issue in hand. You will usually not need any more than three levels to ensure an appropriate degree of operationalisation.

If you use this tool at the project level, you can draw on the conclusions as a basis for operationalisation: Who will do what? When? And with what resources?

If you use it for an area of social concern, the tool will describe sub-processes in greater detail. This will give you a greater insight into the actual situation in the permanent cooperation system, which in turn will allow you to devise specific options for a (future) project.

Tool 27
Process Design

Notes on use

Purpose	This tool will help you develop and specify the details of an individual process.
When to use it	In situations where you need to develop a new (sub-) process or describe an existing process in detail for the first time.
Setting	In small groups with the process participants.
Facilities and materials	Pinboards, flip charts, markers, PowerPoint and video projectors will come in useful.
Notes	You will require a sound knowledge of the processes in question. The tools process map and/or process hierarchy will provide you with a good basis for selecting appropriate processes that require a detailed description.

Description

This tool will help you design and draw up processes in different degrees of detail. Based on an overview of the process to be worked out, you start by describing the entire process before moving on to a more detailed description of the individual steps. You can also define and document individual activities for each step where necessary.

How to proceed

Step 1: Outline the process

Use working aid 31 to outline the new or existing process selected. Here, you define the objectives of the process and specify the process managers, the steps involved and the units or individuals responsible.

Name of the process: Brief description:					
Process manager					
Steps					
Objectives					
Responsible					

Working aid 31: Process design

You should also document additional agreements for the process.

Step 2: Describe the process in detail

In this step, you describe the process in detail and document the following information for the individual activities within the process: the start, duration, dependency on the previous step, responsibility and the number of working days required.

Description of the process					Date:	
Name of the process: Process manager:				Distribution list: Process manager:		
Start: Duration:				Cooperation partner(s):		
Process objective (with indicators):						
Cross-process management/support activities						
No.	Step	Start	Duration	Dependency on the previous step (time, quality)?	Responsible	Working days
1.						
2.						
n						
Process prerequisites:						
Interface with other processes? (Information as required)						
Staff costs:		Travel/workshop costs:		Cost of materials:	Other costs	
Risks/preventive measures:						
Reports (with deadlines):						

Working aid 32: Process definition and process steps

Step 3: Describe steps in detail

Now describe, if necessary, each individual process step within the overall process. To do this, you can adapt the previous working aid and add the following aspects: name of the step, objective, activities, responsibility, cooperation, costs.

Step no. Description of the step					Date:		
Name of the step: Person responsible:					Distribution list: Person responsible for the step:		
Start: Duration:					Cooperation partner(s):		
Objective of the step (with indicators):							
Management/support activities for the step:							
No.	Activity	Start	Duration	Dependency on previous activity	Responsible	Working days	
1.							
2.							
n							
Prerequisites for the step:							
Interface with other steps? (Information as required)							
Staff costs:		Travel/workshop costs:		Cost of materials:		Other costs:	
Risks/preventive measures:							
Reports (with deadlines):							

Working aid 33: Process steps

Step 4: Operationalise the documents

Now use these documents as part of process management, for example, for planning, budgeting and implementation.

Tool 28
Process Optimisation

Notes on use

Purpose	You can use this tool to improve processes and achieve objectives efficiently and effectively.
When to use it	To monitor processes on an ongoing basis using the criteria: ■ efficiency: Can we implement processes faster, at lower cost or achieve higher quality? ■ results: Are the processes still relevant to the project objective or are new processes required?
Setting	In small groups with the actors involved in the process
Facilities and materials	Workshop materials
Notes	To use this tool you will need a good knowledge of the processes. It will prove helpful if you have already prepared a process map.

Description

To optimise processes, you will need to streamline existing processes and focus on a few relevant ones. This means that you not only need to look at how to improve a process but also consider whether the process is really needed, whether it can be dispensed with entirely, or whether it should be combined with another process or outsourced. There are a number of different ways to optimise processes:

Simplifying processes

You can reduce the complexity of processes to eliminate unnecessary loops and allow the process to flow more freely. For example: you can reallocate responsibilities to expedite decisions, simplify administrative operations, and incorporate steering and quality assurance into activities.

Merging processes

You can merge processes in order to reduce the number of interfaces, lighten the coordination workload and lower transaction costs. For example: you can centralise administrative tasks or pool steering responsibilities.

Dispensing with processes

You can dispense with superfluous processes that add no value. For example: you can combine various different meetings into a monthly micro-workshop, disbanding working groups when the context changes.

Changing the sequence of processes

You can change the sequence of processes in order to optimise the production of outputs and reduce transaction costs. For example: you can ensure participation by local stakeholders at an early stage in project planning instead of at the end of the design stage.

Introducing new processes

In order to fill gaps in the value chain you can introduce new processes and/or sub-processes. For example: you can design monitoring and evaluation processes or introduce quality assurance measures.

Standardising processes

You can standardise and simplify processes so as to reduce the need for coordination and communication. For example: you can introduce standardised formats for contracts, lump-sum invoicing, and product design for knowledge products (such as reports, strategies, and evaluations).

Outsourcing processes

You can outsource processes or sub-processes to other businesses or organisations in order to concentrate on the core processes. Support processes – i.e. processes that do not generate any added value in terms of delivering outputs within the cooperation system and can be performed by an external provider without impacting on quality – are ideal for outsourcing. Outsourcing is a particularly attractive option if the processes in question can be carried out more efficiently by an external provider. To do this, however, you must be aware of the true cost of the in-house processes. For example: you could outsource infrastructure maintenance, procurement, training, accounting, etc.

The optimisation options laid out above are designed as an aid for analysing processes. You must decide on a case-by-case basis which option is the most suitable.

How to proceed

Identifying and re-designing processes involves the following four steps. This work should be performed by small teams made up of the actors involved. External support is recommended.

Step 1: Select the process

In this first step, agree on which process(es) is/are to be optimised. To do this, answer the following questions:

- In which processes am I involved?
- Which of these processes do I believe can be improved?

Once the relevant processes have been identified it is a good idea to briefly describe them and explain why you selected them. Answering the following questions may be of help:

- What is the improvement designed to achieve?
- Will optimising processes mean that stakeholders' needs are better met?
- Where and how will the process be optimised?

Step 2: Analyse the process

In this second step, you analyse the process in detail. Answering the following questions and charting the process will help you do this:

- What workflow is used? (Who does what in which order)?
- What is the outcome of the individual process steps?
- What critical interfaces/events will be affected by process optimisation?

The following table is frequently used to depict workflow.

Activities and responsibilities	Necessary resources and support processes	Outcome/product	Individuals affected	Critical event, bottleneck, disruption

- First column: individual tasks performed in full by one person or by an organisational unit
- Second column: resources and support processes needed for the activity
- Third column: outcome or product of the activity
- Fourth column: individuals affected internally or externally
- Fifth column: critical events, bottlenecks and disruptions

Working aid 34: Flow chart

To support process analysis, try using working aid 35 to develop ideas for optimisation.[28]

Critical patterns		Possible solutions
Multi-track, parallel forms of output processes		Merge, network, standardise
A large number of interfaces, processes do not run smoothly, tailbacks at interfaces		Combine duties, one person(or team) sees a task through from beginning to end, staff inter-changeability
Interfaces rather than smooth seams, difficult to connect oper-ations		Reach agreements, strive for client relations based on a spirit of partnership, feedback
Process steps that do not gen-erate any added value, original intention forgotten		Eliminate process steps and do not replace them
Control loops that have fallen into disuse		Remove controls where these are purely formal
Sequential process steps leading to lengthy run times		Perform potentially overlapping tasks in parallel, simultaneous engineering
Many critical steps that waste time and money		Provide support through stand-ardisation, computers, avoid sub-optimisation
Long distances between individ-ual steps in the process (ma-chinery, workplaces), processes arranged by functional criteria		Organise workplaces and equip-ment in line with the structure and design criteria of the output processes
The actors involved are not aware, or not sufficiently aware, of the final outcome of the pro-cess in which they are involved		Make clear the contribution made by actor involved to the finished output or the whole operation. Create process awareness and trust in the process
Inward looking approach, lack of clear allocation of duties, Parkin-son's First Law		Work on the corporate culture, make overarching objectives and results clearer, ensure a clear allocation of duties and authority

Working aid 35: Checklist of critical patterns and weaknesses

Step 3: Define the target process

A lot of ideas for improvement usually emerge during the analysis phase. In this step, you should compile these ideas, flesh them out where necessary and channel them into a target concept (cf. the process design tool).

- Describe the new process and set out responsibilities. Draft a process design.

- Reach agreement about the critical interfaces between process steps.

- Define all steps and activities and have the pertinent decisions taken as required.

- Lay down the targets for the new process (time, costs, quality …).

Step 4: Introduce the improved process

In this step, you define who is responsible for introducing the new process (i.e. who will be the process owner?). Process owners must ensure that the process is documented. They instruct the actors involved in how to use the process and train them if necessary. They are in constant contact with the stakeholders and ensure that the process is enhanced as requirements change. Here are some tips for the introductory phase:

- Inform all those affected in a clear and concise manner.

- Visualise the workflow so that the actors involved have a point of reference if they are unsure about how to proceed.

- Monitor the process closely: Are the objectives being achieved?

- Routinely update process descriptions and check compliance with standards.

- Encourage the establishment of knowledge platforms (e.g. quality circles) in order to eradicate any weaknesses that may emerge.

Tool 29
Interface Management

Notes on use

Purpose	You can use this tool to analyse and optimise interfaces within processes.
When to use it	Any interface is a potential source of error. Therefore, you should review an interface's functionality from time to time, especially when it arises between cooperation partners.
Setting	Small groups, ideally comprising the participants in the process.
Facilities and materials	Pinboards, workshop materials (markers, cards etc.); document handouts.
Notes	Ideally, you will already have compiled the process map and descriptions of the process to be analysed. This tool is closely linked with the process optimisation tool.

Description

At process interfaces, responsibility is transferred from one individual, organisational unit or organisation to another. When optimising processes, increasing efficiency takes top priority. This involves reducing costs while retaining or enhancing the output.

Large number of interfaces: structure

A complex division of labour will result in a large number of interfaces, which in turn will increase the effort required for communication and coordination. Each additional interface increases costs because the actors involved must consult and coordinate their work. This increases the time and costs involved in planning and reaching agreements.

Before you start to analyse the individual interfaces, is will therefore help if you assess whether the number of interfaces is proportional – by casting an eye over the process map.

Poor interface management: process

Losses are often incurred at interfaces. These could take the form of delays, misunderstandings and similar inefficiencies. Optimum steering and interface coordination is a key factor in minimising these losses and boosting efficiency. This tool enables you to determine how the interfaces of an organisation can be optimised.

Here, you examine the following aspects:

- Time: Interfaces can obstruct the process flow or cause superfluous processes. They can pro-long decision-making processes, disrupt communication and duplicate work.

- Quality of outputs: Interfaces can result in quality losses in the outputs produced. The various actors involved may have different opinions on quality; coordination problems may occur when producing the outputs.

- Costs/price: Various aspects of time and quality can generate massive additional costs.

Interfaces in cooperation systems are usually interfaces between organisations. The challenges of working with such interfaces are all part of the logic of cooperation: in the interests of jointly achieving objectives, the organisations involved must sometimes sacrifice part of their autonomy as an organisation. The objective of interface management is therefore to optimise the interfaces to the greatest extent possible, based on the aspects outlined above (time, quality, costs/price), while preserving as much autonomy as possible.

The following basic conditions will come in useful when designing interfaces:

Infrastructure: An adequate degree of communication and coordination is needed. The required infrastructure must be available (e. g. telephone, internet) and joint communication procedures (such as regular meetings) must be in place.

Availability of interface actors: Even the best communication and coordination infrastructure is useless if the relevant actors are not available. The availability of actors is therefore another key condition for engaging in the necessary communication and coordination.

A common language: If they are to coordinate work efficiently at interfaces, the actors involved must speak the same language. In other words, they should have a clear understanding of what they are talking about and how statements made by other actors are intended. A shared under-standing of the objective being pursued by the actors, and how they are pursuing it, is therefore key to interface management.

Assumption of mutual competence: At interfaces, an actor must hand over responsibility for an issue or a process to another actor. To be able to do this, the actor in question must have confidence in the abilities and capabilities of the receiving party. This trust must be built and maintained.

How to proceed

Step 1: Identify interfaces and recognise problems

Before you can optimise interfaces, you must first determine where the interfaces lie in the process to be examined. Once you have done this, you must pinpoint the ones where problems were identified.

Step 2: Analyse interface problems

Use working aid 36 below to determine the possible reasons for the problems.

General conditions	Key questions
Infrastructure	Is it physically possible to maintain the necessary flow of communication? (telephone, internet, etc.) Are the individuals/organisational units/organisations affected sufficiently familiar with how and when they can and should use the existing communication infrastructure? Are mechanisms in place to safeguard the flow of information? (such as routine meetings and reports)
Availability of the individuals affected	Is there always somebody at the interface? Are the individuals responsible for coordinating the interface available in principle and can they be contacted? Is there a constant fluctuation in contacts or do you always deal with one person for interface coordination?
Standardised communication/ common language	Do the individuals involved in the coordination process speak the same language? Are they pursuing the same objectives in terms of the overall process? Do they aim to take the same path to achieve these objectives?
Accepted capabilities	Are there structures that make it more difficult to ensure interface cooperation on an equitable basis? Do the relevant interface actors trust each other? Do they acknowledge one another's capabilities? Is the distribution of capabilities in keeping with the structure for the overall process?

Working aid 36: Checklist for analysing interface problems

Step 3: Discuss possible solutions and document conclusions

Based on the findings of the analysis carried out in step 2, you now discuss possible solutions. You can use working aid 37 to document the conclusions of this discussion.

Interface:			
Description of the interface	Individuals/ organisations involved	Problems with interface	Solutions

Working aid 37: Interface management

Step 4: Implement solutions

It is often expedient to assign the mandate to improve the interface problem to the relevant process owner.

Success factor Learning & innovation

Tool 30
Scaling-Up

Notes on use

Purpose	You can use this tool to scale up and mainstream experience, learning process-es, knowledge and solutions to ensure that innovative pilot activities and new approaches generate broad-based, structure-building impacts.
When to use it	For scaling up the design of innovative pilot projects and new approaches with a strategic orientation. Use it to replicate and/or disseminate successfully implemented or tried-and-tested innovations and change processes.
Setting	You can conduct discussions within the scope of strategy/concept development or during planning or evaluation; you should include key actors.
Facilities and materials	Printouts or visualisation of the checklist, depending on the setting.
Notes	Start with a description of the innovation to the scaled up that is as precise as possible. To scale up innovative approaches you will need your own strategy development process, which this tool can support.

Description

One key issue that arises when dealing with change processes in any area of social concern is how you can achieve results that are as broad-based as possible. Scaling-up is one way of replicating innovative, tried-and-tested approaches on a wider scale. Taking the learning cycle of variation, selection and restabilisation as a basis, innovative approaches often emerge as part of variation, i.e. through minor or major modifications to the established routines in the area of social concern. These variations can prove invaluable when it comes to identifying innovative, tried-and-tested approaches that can be scaled up. In order to ensure that you make the right choice (during selection) the approaches you identify should have the potential and the capacities (in terms of combining and coordinating political will, interests, knowledge, values and financial resources) to be replicated. In the relevant area of social concern, scaling-up ultimately involves describing the innovative approach in relevant standards and manuals and devising a capacity development strategy for the replication process (restabilisation). You can choose to do this by replicating pilot approaches step-by-step (a process referred to as horizontal scaling-up) or by mainstreaming new concepts in laws, strategies and policies (vertical scaling-up). When devising innovative approaches and pilot activities, it is important that you take the potential for scaling up into account from the very start.

The following key factors play an important role in achieving broad-based results:

Key factors	Activities
Incorporate scaling-up into planning	Take scaling-up into account from the outset: Set goals, identify actors, assess capacities and risks and explore financing concepts.
Ownership by the key actors	Key actors undertake to achieve broad-based results and secure political backing.
Multi-level approach	Link up policy advice with implementation models in selected local or regional application contexts.
Verification of results	Substantiate innovative approaches within the framework of monitoring.
Standards and manuals	Describe the necessary process steps and tools in manuals and standards.
Replication structures and incentive mechanisms	Institutions must be able to shape change processes (training and organisational development). Develop incentive and replication mechanisms that extend above and beyond regulatory approaches.
Communication and networking	Incorporate key actors and population groups through information, communication and networking.
Scheduling and budgeting	Earmark adequate financing and schedule sufficient time.

Figure 42: Key factors in scaling-up

How to proceed

Step 1: Select the item to be scaled up

Which innovation do we wish to scale up and mainstream? Start off by obtaining a precise and concise description of the approach that has been selected for scaling up:

- What practices have proven their worth in terms of effectiveness? e.g. goals, risks, standards, manuals
- Can the practice be generalised and applied to another context? e.g. at the cultural, social, financial and institutional level

The following checklist will help you identify the issues you need to address in the scaling-up process.

Steps in the process	Key questions	++	+	–	– –
Evaluating experiences	Are the innovative approaches and best practices described in a precise and structured manner?				
	Are you sufficiently familiar with the financial and institutional conditions for scaling-up?				
	Are the rights regulated so that third parties can use and further develop any innovations that arise throughout the course of the project?				
Actors	Have you conducted an analysis of actors and discussed it with different actors?				
	Do the key actors possess the capacities required for scaling-up?				
Scaling-up strategy	Have you discussed the scaling-up hypotheses with the key actors?				
	Have you discussed and agreed on milestones and cut-off points with the relevant actors?				
	Have you discussed various options for scaling up and reached a sound decision in favour of one of them?				
Resources	Are there sufficient human and financial resources for the start-up phase?				
	Are there sufficient financial resources in place for scaling-up, or is there a financing model?				
Monitoring and quality assurance	Do the relevant actors have access to existing tools and structures to observe and steer the scaling-up process?				
	Are the relevant actors familiar with the core of the innovation to be scaled up?				

Working aid 38: Checklist for scaling-up

Answering the questions in the checklist will help you devise lines of action that you can incorporate into strategy development and formulation.

Step 2: Formulate a strategy

When designing a scaling-up strategy, you need to highlight key aspects and present them in an appropriate form. The tool 'Strategy suite' in the success factor 'Strategy' describes strategy development in detail. The following questions provide a rough outline of the aspects you need to examine in greater detail in this context:

- In which social context will scaling-up take place?

- What actors will it affect? What will be their role? What degree of ownership will they have? What actors need to be involved and informed? How will this happen?

- What capacities are required and available? What incentives and financial options are there?

- What options for scaling-up are conceivable? What criteria should be used to select a suitable option?

- What results are to be achieved? (results model and hypotheses)

Channel the answers to these questions into a capacity development strategy that will form the basis for a plan of operations.

Step 3: Provide resources

In this step, you calculate and negotiate the financial and human resources that the actors involved will need to disseminate and sustainably mainstream the strategy.

Step 4: Implement scaling-up

Here, you implement the scaling-up process in line with the capacity development strategy. New actors may influence the process, in which case you will need to engage in negotiation processes again. Joint monitoring of the results and of the plan of operations safeguards quality assurance and steering of the process.

Tool 31
Learning Capacities in Cooperation Systems

Notes on use

Purpose	You can use this tool to review the learning capacity of a cooperation system. In other words, what resources does the system have to proactively assimilate inputs and continue developing?
When to use it	In situations where you need to strengthen the sustainability of a cooperation system.
Setting	Workshop with key actors.
Facilities and materials	Form to assess the learning capacity (as a document or visualised on a pin-board).
Notes	You will need a sound knowledge of the cooperation system and its actors.

Description

The more a cooperation system has fixed structures, processes, rules and rituals, the more targeted and efficiently it can act. The greater the degree to which cooperation systems can use the mechanisms of variation, selection and stabilisation, the more sustainably they will achieve their strategic objectives and results.

How to proceed

Step 1: Define clearly the cooperation system's boundaries

Start by identifying the boundaries of the cooperation system as the object of observation and evaluation. Boundaries that are drawn too widely can rapidly result in inaccurate negative judgements. If they are drawn too narrowly, some dimensions of observation may perhaps appear more positive than is actually the case.

Step 2: Rate learning capacity based on questions

The following form proposes seven key factors for rating the learning capacity of a cooperation system. The more positive the rating, the greater the learning capacity.

Factors of learning capacity	Questions	++	+	−	−−
Legitimacy of the cooperation system	How committed are the cooperation system's actors to participating in the joint change processes (stabilisation)?				
	How clear are the objectives to the participating actors? How willing are they to take on responsibility for achieving these objectives? (selection)				
Investment of resources	Do the actors contribute new ideas and initiatives for activities? (variation)				
	Do ideas and initiatives come from different actors? (variation)				
	Do the actors involved provide sufficient resources for the cooperation system? In other words, have individuals who feel responsible for involvement been appointed? (stabilisation)				
Response to problems	Do different actors identify obstacles that jeopardise the achievement of objectives and address these in subsequent activities? (variation)				
	Does the cooperation system address new challenges and solutions? (selection)				
Communication	How binding and reliable is communication between the actors within the cooperation system? (stabilisation)				
	Have rules of communication been established and formalised? Are they being observed? (stabilisation)				
Cooperation	Are the chosen forms of cooperation appropriate for high-quality cooperation? (variation, selection, stabilisation)				
	Are the rules of cooperation established and documented? Are they being observed? (stabilisation)				
	Do the participating actors invest adequately in the trust factor? (selection, stabilisation)				

Factors of learning capacity	Questions	++	+	–	– –
System boundaries	Is there a balance between cohesion and openness in the cooperation system? In other words, is the system cohesive yet at the same time open to new actors? (variation, selection)				
	Is the search for and integration of new actors oriented to the strategic objectives? (variation, selection)				
Quality management	Are structures, processes, rules and rituals within the cooperation system periodically addressed and adjusted with a view to achieving objectives, where appropriate? (variation, selection)				

Working aid 39: Checklist for learning capacity of cooperation systems

Step 3: Define activities to strengthen learning capacity

Enter the strengths and weaknesses of the cooperation system as regards learning capacity in the form. This will allow you to identify suitable approaches and activities for fostering learning capacity.

Tool 32

Innovative Capacity of Cooperation Systems

Notes on use

Purpose	Use this tool to assess the innovativeness of a cooperation system.[29]
When to use it	Where you need to develop the innovative capacity of the cooperation system.
Setting	Suitable as a checklist for rapid assessment and as a basis for in-depth discussion by the key actors.
Facilities and materials	Copies of the evaluation table; it may be appropriate to display the table on a pinboard, and to work with dots.
Notes	You will need a good knowledge of the cooperation system. When working in a large group, make sure there is practical scope for evaluating individual assessments. You can also apply the tool to individual organisations.

Description

Innovative capacity is the ability of individuals, groups, organisations, cooperation partners and networks to drive innovations. This involves a lot more than just coming up with new ideas. After all, ideas only generate innovations if they are implemented in the form of new products, services or procedures that are actually used at a practical level.

Extensive empirical studies have been published on what factors are conducive to driving the capacity for innovation. The answers they provide are very diverse indeed. For example, interaction between the following factors is frequently deemed to have a positive influence on innovative capacities:

- ongoing competence development;

- holistic innovation management;

- a work culture that is conducive to learning (a tolerance for errors);

- innovative forms of work organisation such as teamwork, shorter information paths;

- integration and appreciation of diversity (diversity management).

You start the process by reflecting on the ability of a cooperation system to implement the three learning mechanisms for a cooperation system (variation, selection, (re)stabilisation) before examining the cooperation system's innovative capacity on this basis, using the following four criteria:

- culture of innovation;
- strategy for innovation;
- resources for innovation;
- structures for innovation.

How to proceed

Step 1: Discuss general points

Start by discussing the questions outlined below. This will help you prepare for a detailed innovation check, based on the three learning mechanisms (variation, selection, restabilisation).

- How variation-friendly is the cooperation system? How easy or difficult is it to allow variations in routine operations within the cooperation system?
- How selection-friendly is the cooperation system? Is the cooperation system able to make decisions that will allow innovative variations to be developed and implemented?
- How stabilisation-friendly is the cooperation system? To what degree is the system able to sustainably mainstream innovative changes in routine operations at the product, service and procedural level?

Detailed reflection on these points will usually help provide a sharper focus for the subsequent assessment.

Step 2: Assess the innovative capacity

In this step, you assess the innovative capacity of the cooperation system by rating four criteria using three statements. You then rate the criteria from 0 (strongly disagree) to 5 (strongly agree). Working aid 40 provides an example of such a rating.

Criterion	Key statements	Rating (1–5) for example
Culture of innovation	Innovations enjoy high status within the value system.	3
	A positive culture of errors is in place.	1
	There are concrete incentives for innovative thinking and action: acknowledgement, reward etc.	4
Total		8

Criterion	Key statements	Rating (1–5) *for example*
Strategy for innovation	The thematic areas and lines of action for innovation are clearly defined and communicated.	5
	The learning mechanisms that promote innovative capacities are clearly defined.	2
	Strategic and operative bodies are appointed that represent a wide variety of perspectives.	4
Total		**11**
Resources for innovation	Funding is available for innovations.	0
	The individuals involved have strong innovative capacities.	3
	The organisations involved have strong innovative capacities.	2
Total		**5**
Structures for innovation	Communication structures are established on the basis of functionality, transparency and diversity.	4
	The cooperation system includes structural elements and scope for creative thinking (e. g. learning workshops and learning journeys).	4
	Relationships with external actors are shaped in order to build innovative capacities.	4
Total		**12**

Working aid 40: Table for rating innovative capacities

Step 3: Interpret the findings

In this step, you map the findings in a spider chart, as shown below. You can visualise both the ideal innovative capacity and the actual innovative capacity of the cooperation system. The chart is easy to use and allows you to review the strengths and weakness of the innovative capacity and to interpret the findings at a glance.

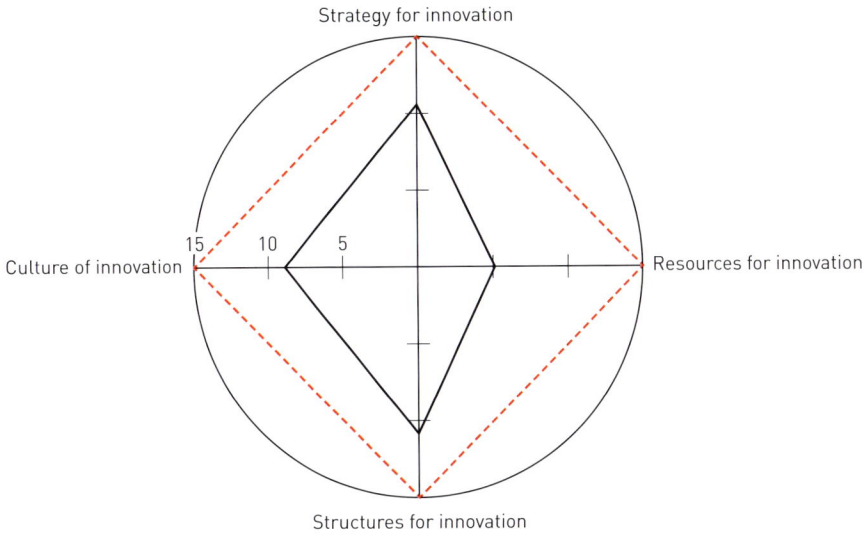

Figure 43: Example of an assessment of the innovative capacity

Possible interpretation of the innovative capacity of the cooperation system in the above example:

Although the cooperation system would like to be innovative, as manifested in the innovative aspects of its strategy and structures, insufficient resources are being made available. This could also be the reason why the culture of innovation is weak. The system talks about innovation, but doesn't practice it because this would incur costs. This means that there is a lack of incentives for the cooperation partners: innovation for them means having to make more effort, with little motivation for creative thinking and action.

Step 4: Define activities

In this final step, you agree on activities that will build the innovative capacity of the cooperation system.

Tool 33
Knowledge Management in Projects

Notes on use

Purpose	This tool will help you identify and safeguard relevant practical knowledge, and systematically manage it for sharing with others later on.
When to use it	For defining knowledge management goals (during annual planning, which also includes allocating resources), and for creating knowledge products
Setting	Workshop with relevant actors.
Facilities and materials	Project documents that support the recording of key lessons learned.
Notes	The tool focuses on creating knowledge products rather than on systematically setting up a knowledge management system. Realistically speaking, you should concentrate on no more than two issues in the project and analyse and document them from the perspective of the knowledge product's subsequent beneficiaries.

Description

Learning and innovation are based on knowledge – knowledge about the project, about individuals and organisations involved and their environment. This knowledge arises at various points within a project and is usually made available in a condensed and explicit form, for instance as a strategy, concept, description of a procedure or problem-solving path, or a method or tool.

Knowledge management at the project level serves two purposes: On the one hand, the lessons learned and experiences gained in projects are a driving force for joint learning among the actors involved. On the other, it feeds the knowledge that arises within projects back into the organisations involved and gets them – as the actors in the cooperation system – into shape for sustainably mainstreaming innovations over the long term.

Knowledge management will help you answer the following key questions:

- Which strategies and procedures have proven their worth in practice, and have contributed to effective cooperation? What can the project learn from them? What can others learn?

- What knowledge will also be of benefit to the actors involved in the future? What knowledge has proved useful? What knowledge is no longer relevant?

At the project level, the aim is to identify, define and disseminate relevant knowledge together with the participating actors. In the process of cooperation the participating actors develop and apply a variety of knowledge products, for example:

- (a) a **strategy** (project concept, guideline, rule, policy) that provides orientation and has proven effective and economically efficient;

- (b) an **explanatory model** that helps us better understand, represent and interpret a situation;

- (c) a **problem-solving path** that shows how a problem can gradually be solved;

- (d) a **learning history** that shows how the project learned from its mistakes and how mistakes can be avoided in the future;

- (e) a **description of a change process** in which the participating actors were able to negotiate their different interests and achieve their objective;

- (f) a **tool** or method that helps generate a response to a specific question and leads to a tangible result;

- (g) methods for representing and transferring **knowledge** to other contexts.

Useful knowledge products:

- speed up and simplify procedures and are replicable, provided that they are made available in an appropriate form and adapted to the given context.

- facilitate cooperation within the project and boost an individual actor's performance capability;

- highlight 'blind spots' within the project, open up access to external sources of knowledge and condense externally acquired knowledge into a comprehensible form;

- are presented simply, comprehensibly and concisely;

- sharpen your awareness of unutilised potentials and learning needs;

- give the organisations involved a clear profile that can be used for instance to share information with other bilateral and multilateral organisations.

Depending on the target group, you should use 'traditional' formats such as brochures and flyers to make knowledge available. Alternatively, you can use web-based tools such as blogs, webinars and communities of practice to exchange information. Online tools allow you to update content quickly. Comprehensibility and transparency are supremely important. A mixture of text, images, graphics and concise stories/examples plus a suitable mix of media facilitate exchange and dissemination.

How to proceed

Step 1: Plan knowledge management goals

Knowledge management is a task for all actors in the project. It therefore makes sense if the actors involved in a project ask themselves when drawing up annual plans what knowledge products are to be produced.

- What knowledge do you and your cooperation partners wish to analyse and document for each other?

- What knowledge do you wish to analyse and document for the project?

- What knowledge do you wish to analyse and document for sharing with others?

Step 2: Identify areas to be looked at

Projects always reflect regularly on their work (through strategic controlling, monitoring and evaluation) in order to learn from their experiences. During annual planning or when reviewing evaluations, agree on knowledge management goals. In this context, identify the areas to be monitored for which a knowledge product will be produced:

- What issues and problems were the focus of attention the previous year?

- Where did you develop a solution that could also be useful to others?

- Where did the selected approach fail? What can you learn from this for the future?

- Who might be interested in that other than the actors in the project? Who should receive this information?

Step 3: Define the focus of knowledge products

In this step, focus on the following criteria:

- the possible benefits generated by the product, for instance as a result of it simplifying or speeding up a procedure;

- the product's target group;

- the distinctiveness and profile of the product, which mean that its features can be communicated to others;

- the degree of innovation, which enhances the distinctiveness and profile of the product;

- the scope for supporting users.

Step 4: Create knowledge products

Knowledge management products can be created by the actors involved in the project or by external consultants and experts. They can be developed by individuals (e.g. reports on lessons learned) or in workshops (lessons-learned workshop) involving the individuals who acquired the relevant knowledge. They can be developed in network structures, i.e. in flat, self-organised and open communities of practice or as part of standardised procedures (such as debriefing).

You can use the following working aid to prepare and describe a knowledge product:

Title of the knowledge product (use a catchy title where possible)	Brief description (logical, simple, comprehensible, practical)
Theme and context ■ What is the product for? ■ What does an interested individual need to know about the context?	
Description of content ■ What issue does the product address? ■ How did you proceed? With whom? ■ What was especially helpful in that context? What was the secret of this success? ■ What stumbling blocks or obstacles did you encounter? How did you overcome these? ■ What risks should people look out for? ■ What minimum requirements must be met? ■ In which other contexts would the product be suitable?	
Benefits and results ■ Who will find the knowledge product useful? ■ What are the intended results of the knowledge product? ■ What was innovative, new and unfamiliar about these results? ■ What is the estimated cost of the application? How much effort will be involved?	
Contact and support ■ Who is available to provide further information? ■ Who will support the users?	

Working aid 41: Producing and describing knowledge products

Step 5: Share and disseminate knowledge products

You can now share and disseminate the products through various channels:

■ via peer groups;

■ via internet and intranet platforms;

■ at forums, congresses and sector networks;

■ as printed publications;

- in communities of practice, microblogs or similar, web-based shared learning platforms;

- informally (e. g. in the cafeteria …)

The focus here is not just on exchanging information with other actors and on scaling up the products outside the project, however. This step will also help you disseminate the lessons learned among the actors involved in the project itself. How will they assimilate the lessons learned and mainstream them within their organisations and within the permanent cooperation system? Generating dialogue about products will not only open up new opportunities for mutual learning and for sharing existing knowledge, it will also give rise to new knowledge.

Tool 34
Debriefing

Notes on use

Purpose	This tool will help you learn lessons from successes and problems and safe-guard this knowledge for the benefit of the project and other future projects.[30]
When to use it	To evaluate a completed project, and for structured reflection when new partici-pants come on board and following particular events or milestones.
Setting	Workshop with the relevant participants or structured discussion.
Facilities and materials	Pinboards, workshop materials (markers, cards, etc.).
Notes	It is a good idea for you to prepare well for presentation of the project trajectory. The moderator must ensure that the process does not involve disparaging re-marks, the attribution of blame or exaggeration of positive aspects.

Description

You gain valuable experience throughout the course of a project. Some things succeed, while in other areas you make mistakes and problems occur. And as everyone knows, you learn most from your own experience. Debriefing is a dynamic, joint learning process that is based on the lessons learned gleaned during a review by the actors involved.

It generates a wide variety of benefits, including:

- You can replicate successful aspects.

- You can avoid mistakes you have already made.

- You will be better able to anticipate resource shortages and areas of risk.

- You can optimise standard processes.

- The quality of planning is improved in the light of experience.

- Individual experiences can be combined to produce new group competences and joint lessons learned.

- New participants benefit from lessons learned.

- You can use standardised procedures for handover processes.

- A productive, shared culture of learning arises in which mistakes and problems are seen as providing opportunities for improvement.

Although a debriefing workshop does not create any excessive demands with regard to preparation or infrastructure, there are a number of obstacles and stumbling blocks that you should take into account so that you really can learn long-term lessons from a project.

- It is important that you get the wording of the invitation to attend debriefing right, as this influences participants' attitudes. If you describe a debriefing workshop as a project evaluation or as an assessment of participants, this will not pave the way for openly discussing mistakes and problems.

- The quality of a debriefing workshop depends first and foremost on the quality of the moderation. The person responsible should be as neutral, impartial and experienced as possible. This means that this role should only be assumed by a project participant in exceptional cases.

- If a workshop is poorly prepared or moderated, debriefing can turn into superficial flattery, or into a situation where participants are blamed for outcomes, which can impact on motivation. To ensure that you achieve the aforementioned benefits and steer clear of any risks, you need to use a tried-and-tested method and ensure that the individuals involved are open to taking part in the process.

- If a debriefing workshop takes place a long time after the experience, the particular lesson learned may no longer be fresh in the minds of the individuals involved, or they may be less motivated to discuss it. This also makes it more difficult to directly put into practice the lessons learned. We recommend that you carry out debriefing workshops shortly before someone leaves the project or that you integrate it into project planning, ideally directly after key milestones have been achieved in projects/sub-projects.

- Depending on the size of the project, we recommend that you include all actors who are involved in the project early on in the debriefing process in order to be able to integrate as many perspectives on things as possible.

How to proceed

Step 1: Establish why the debriefing workshop is being carried out and identify the issue to be examined

The scope and structure of a debriefing workshop will depend on the reason why it is being carried out and the issue to be examined. Therefore, you should clearly state the benefits of debriefing in the invitation, so that all participants are clear about its purpose. There may be a particular issue that the workshop needs to address. The more clearly you communicate the reason for the workshop and the issue it will examine, the more productive the debriefing process will be.

Step 2: Collect successes and problems

The approach you choose to adopt will depend on the time available and the complexity of the project and the group involved.

The individual participants recall successes and problems that occurred in the course of the project and each participant then writes these down on cards. In complex projects, you can carry out debriefing for parts of the project (e. g. lines of action, work packages).

It is helpful if you map the cards on a timeline of the project's key milestones by having participants position them under the corresponding milestone. When they pin up their cards, you should give each participant an opportunity to comment on the successes and/or problems in front of the plenary group or to pin them up without comment, or hand them to the moderator anonymously for him/her to pin up.

Alternatively, have the moderator collect the participants' feedback by email prior to the workshop. This step generates a revealing map of successes and problems that already highlights critical phases or events at a purely visual level. It also reveals whether the participants perceived the same events as key successes or problems.

Experience has shown that it is worthwhile gathering the successes and problems on an open-ended basis. Alternatively, you can collect the information in a structured manner using the following questions.

- How did the **objectives system and the results model** look at the start? How did they change over time?

- What has been achieved? What are the **results**? What key moments were there? What was the biggest milestone achieved?

- What were **successful steps** or strategies with respect to the achievement of objectives? Which steps or strategies should be continued?

- In hindsight, what could have gone better in terms of the achievement of objectives? What has proved unsuccessful? What made the **achievement of objectives** difficult or **impossible**?

Step 3: Cluster successes and problems and define lessons learned

In this step, the moderator pools the successes and problems into similar categories together with the participants. This can be done for instance by asking the following questions:

- What do the successes or problems have in common?

- How did these successes or problems occur?

Participants visualise their observations and hypotheses and elaborate lessons learned.

- What general lessons can we learn from these successes and problems?

- What individuals and organisations played an important role in the successes and problems?

- How can you ensure that similar projects will run better?

- Have lessons been learned that would be particularly suitable as examples of best practice?

If particularly serious problems have been identified, these can be analysed in greater detail. This in-depth debriefing method involves jointly reconstructing a decision-making process: 'Who/what body made a particular decision? When was this decision made and what was it?' This helps you to identify key decisions that subsequently turned out to be wrong or less than optimal, so that you can avoid them in the future.

Step 4: Draw conclusions and define activities

In this step, you ask yourself what approach the project needs to take in order to meet the needs of the objectives system. The aspects discussed in the previous steps allow you to document recommendations for the project's steering (structure) in line with the five success factors. It is useful if you map these recommendations on a pinboard, focusing on the objectives and results. Try to cover all five success factors.

Based on the outcomes achieved so far, you now invite all participants to discuss and develop activities and recommendations.

To support you in carrying out this step we recommend that you gather any project documents on objectives and results (such as the results model) or on the success factors (e. g. map of actors, the process map, the capacity development strategy, strategic options, the steering structure, and the architecture of learning) prior to the debriefing workshop and that you make them available for the discussion. You can include these documents in any documents handed over following the workshop.

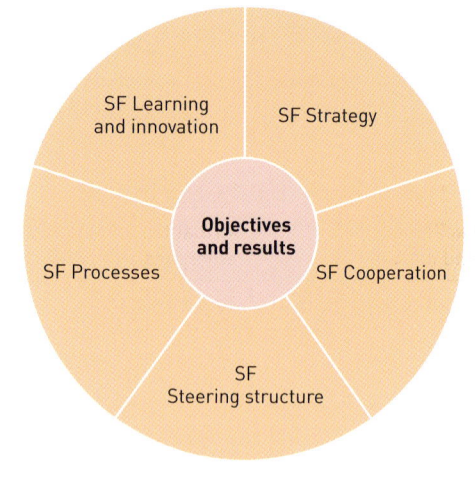

Working aid 42: Debriefing – visualising activities

Step 5: Document lessons learned

In this step, the moderator documents the lessons learned and hands them over to the project. They serve as a basis for deciding on the project's future orientation and for learning from the experiences, successes and problems of the past.

The following formats have proven appropriate and effective for documenting the lessons learned:

- micro articles (short, lively articles on projects);
- case studies (contextually rich project chronologies);
- learning histories and project maps (graphic methods for visualising project trajectories);
- knowledge products as structured formats for knowledge management (for more information, see the tool 'Knowledge management in projects').

Tool 35

Learning Networks for Multipliers and Trainers

Notes on use

Purpose	This tool will help you set up and structure networks of multipliers and trainers to exchange examples of best practice, generate new knowledge and build proactive capacities.
When to use it	Suitable for networking institutions that wish to exchange information on methodological, didactic and sector-specific issues. Can also be used transnationally. Particularly suited to making training material widely available and safeguarding competence building in basic and further training institutions. Supports the sustainable strengthening of capacity-to-build-capacity and cross-institutional knowledge sharing.
Setting	Depending on requirements, this tool can be used in the different phases of networking and knowledge production.
Facilities and materials	Virtual tools that enable the co-creation of knowledge products (such as wikis, etherpads and other read-write tools).
Notes	To use this tool, you will need a sufficient number of organisations that are interested in exchanging information on methodological, didactic and sector-specific issues and on knowledge creation in general. These organisations should not be direct competitors.

Description

Learning networks for multipliers and trainers enable participants to learn with and from each other within the context of participatory, equitable 'knowledge cooperation' (peer-to-peer learning). They are an important knowledge resource that can enrich individual projects. In this way, they can help strengthen capacities in a particular area of social concern, in line with the capacity-to-build-capacity approach (for instance by strengthening the competences of further training institutions which are in turn able to provide skilled trainers to support the target groups of projects in the long term).

When using this tool, it is important that you remember the following golden rules:

- Knowledge must be put into context, i.e. it must be localised to real-life situations.

- Knowledge transfer involves more than just passing on know-how for example by providing multipliers and trainers.

- Knowledge arises in a collective learning process that involves different groups (of experts).

Knowledge that is created collectively is usually owned by the group that generated it. As a commodity that is governed by the law of the 'commons', it can therefore be made available for further use to the group and possibly to other interested parties at no cost using open-source 'licenses'. This makes it easier for knowledge products to be updated locally and distributed at no cost, enabling the development of innovative business models on the ground.

The co-production of knowledge focuses first and foremost on involving learners as experts with a wide array of know-how and experience. As such, it constitutes a collaborative learning process in itself that revolves around ownership and the independent creation of knowledge by a learning network.

Learning networks of multipliers and trainers and communities of practice share certain traits, such as a 'culture of freedom'. However, they focus explicitly on optimising multiplier and training processes.

How to proceed

Step 1: Define the learning network's goals

Even if reaching consensus constitutes a key structural element of learning networks, you still need to clearly define the goal from the outset. This goal will span the methodological, didactic and/or sector-specific needs of the local further training institutions and units, for example, along with those of the sectoral target group(s) they serve.

Step 2: Initiate peer-to-peer learning processes

Learning networks evolve when participants undergo learning processes with others. To do this, you need to identify suitable forms of learning support that address the needs that arise in order to achieve the defined objective and structure the acquisition of appropriate knowledge. Learning support organises and moderates the joint knowledge production process, for example by having the participants create training materials for a train-the-trainer cycle.

The next part of this step involves defining selection criteria for the participating peer learners. Here, particular emphasis is placed on the scope of influence of the individuals involved and of their 'home' institutions in order to pave the way for replicating the learning process and the findings from the outset. On the one hand, this institutional mainstreaming provides the 'raison d'être' for the sustainability of the learning process and of the network itself. Creating an enabling environment within the participating institutions (e.g. by linking specific functions and job profiles with participation in the learning network) will ensure that collaboration continues despite staff turnover or job rotation.

Select applicants in an open call for participants designed to attract the most suitable individuals and institutions, including those with which you were not previously familiar.

Step 3: Conduct peer-to-peer learning event

You usually conduct peer-to-peer learning as a mix of online and face-to-face phases (e. g. online preparation and preliminary networking, face-to-face train-the-trainer sessions, online follow-up to build communities). To ensure that knowledge products are created collaboratively, make sure that participants are involved in all of the development phases and in fine-tuning the materials and products. Throughout all phases, the emphasis is on:

- exchanging experience and perspectives on an equitable basis;
- strengthening replication capacities and broad-based scaling-up at the local level to ensure that knowledge is disseminated beyond individuals;
- transferring methodological competence (such as peer-to-peer-based adult education, moderation methods, competence for change);
- the joint contextualisation of lessons learned (for instance, developing context-specific curricula, trainer's handbooks and training materials).

The co-production of knowledge is driven by the following key core groups:

- content community: this group of experts has a background in the sector in question and is responsible for preparing the training content (especially in the kick-off phase)
- trainer community: a group of multipliers and trainers who have a knowledge of the specific local context and is responsible for the methodological and didactic structure. Representatives of this group are familiar with running training courses in the relevant area of social concern. They form the core of the peer-to-peer learning network.
- community of practice of practitioners and training participants: an extended circle of practitioners and training participants who update and evaluate the materials.

Knowledge production is based on the principle of constant input and feedback among these three core groups. The training materials produced in this way are geared to the local context and can be made universally available. When the materials are later used in training activities the learners' experience is harnessed in order to supplement and tweak the content, and thus further develop the materials. This leads to the emergence of a self-organised, self-perpetuating process of peer learning that supports sustainable knowledge sharing through the continuous updating of training materials and other processes. New business models are also created (for modifying and selling the open source training materials and corresponding services for instance) for local multipliers and trainers as well as for further training institutions.

Step 4: Enable self-sustaining peer-to-peer exchange

Tools that enable peer-to-peer exchange must be made available in order to establish a sustainable learning network. Web-based community-building tools will help you set up a learning community (such as a trainer community) and will allow participants to exchange information and experience once the event itself has taken place.

Establishing product and service development platforms can also help you to develop and sell products and services for a learning network. You could also use these platforms for reporting and monitoring activities carried out by the participants such as independently run training measures and follow-up activities carried out in projects. E-coaching measures will not only help you to safeguard quality, they also round off activities to follow up on the establishment of learning networks.

Tool 36
Communities of Practice

Notes on use

Purpose	This tool is a non-hierarchical, practical form of learning for sharing knowledge and experience. Individuals with shared interests exchange information on a defined area of specialisation and generate new knowledge together.[31]
When to use it	Highly suitable for mobilising implicit knowledge and using practical experience for further learning.
Setting	Depends on your specific objectives.
Facilities and materials	Communities of practice usually operate on web-based platforms for learning and exchange. You could also use a moderation format, however, if the individuals involved are able to meet in a particular location.
Notes	A sufficient number of individuals who are working on the same issue and are interested in exchanging relevant ideas and experiences is a prerequisite. It is crucial that you define clear objectives, identify the relevant roles and appoint individuals to carry them out, and where required secure the backing of management. You can use the tool at the organisational level as well as in cooperation systems. Communities of practice can be used to exchange specialised information at the local and regional right up to the global levels. Establishing a community of practice is time-consuming, and the effort required should not be underestimated.

Description

A community of practice (CoP) is a group of individuals who share an interest in a common field and join forces to actively exchange practical knowledge and experience over a long period of time and to generate new knowledge together. Participation is voluntary and cannot be delegated. CoPs trigger collective learning processes that generate knowledge and experience that is continuously developing. They are extremely effective, help develop an institutional memory and generate new knowledge and skills that are channelled into cooperation systems and organisations. CoPs link up practitioners in a manner that transcends the boundaries of organisations (and of organisational units or countries), irrespective of the hierarchical position that these practitioners hold.

CoPs share the following basic features:

- **Needs orientation:** CoPs come into being and continue to exist because of the shared needs of members. Needs cannot be enforced on members – they themselves determine the priorities.

- **Practical orientation:** They have high regard for practical solutions that should address members' specific needs and be easy to implement in their daily routine.

- **Learning orientation:** Members are interested in the experience of others, because they assume that the latter find themselves in a similar situation and can help develop appropriate approaches to achieve an optimal outcome. Sharing knowledge and providing mutual support generate open and creative learning processes.

- **Respectful communication:** CoPs are successful when communication between members is characterised by fairness, solidarity, transparency and openness. Participation and equitable, non-hierarchical communication help boost ownership and self-organisation.

CoPs are frequently driven by the individuals who are responsible for a particular task and know best what is required to carry it out more efficiently and to deal with influential factors in the immediate environment. These individuals are interested in finding out how others approach similar tasks and what solutions have already been found to successfully tackle a particular problem.

Highly formal organisations with a rigid hierarchy tend to trust expert analyses rather than the practical knowledge of their own staff members and the hands-on lessons they have learned. The same applies to cooperation systems in hierarchical and highly formal environments. Communities of practice offer an efficient and effective alternative. Members exchange information by looking over one another's shoulders. This plays a key role in generating knowledge, promoting learning and fostering development and innovation.

CoPs develop their own momentum and steer themselves as a group of like-minded actors. This facilitates the ongoing process of consolidating the capabilities of actors, creating an environment of transparency and trust.

Alumni networks are one particular type of CoP and bring together individuals who have shared a formative learning experience. They can help safeguard the continuity of learning and the sustainability of results even after the project has finished. Linking up alumni networks with international expert groups can help give rise to innovative forms of cooperation (such as web-based learning networks) that generate a particular added value both for the individuals involved and in terms of further developing social issues.

How to proceed

Step 1: Define the community's objectives

In this first step, you agree on the objectives together with the other individuals whose idea it is to set up a CoP. These could include:

- exchanging information on a specific issue;
- developing potential solutions;
- testing practical applications;
- introducing a change process to mainstream the solutions.

Depending on the issue, decide which other participants you may need to approach because they play a key role in addressing the particular issue.

Step 2: Define the roles in the CoP

After a certain amount of time, the dynamic between the different groups of individuals involved in the CoP stabilises as they form ties of varying strength within the community. We normally distinguish between the following groups:

Core group: This is the main hub of the CoP and normally comprises the community's organisers and moderators. The core group makes sure that there is a healthy balance between freedom, observation, support and requests for contributions within the group. Experience shows that a CoP needs a moderator to sustain momentum. Otherwise, activities will come to a standstill.

Inner circle: This is an informally structured group of individuals that meet regularly (online or face-to-face) to exchange information.

Outer circle: The members of this group rarely make active contributions of their own. They are included in the email distribution list and although they have write access to the online platform, they may choose to just read others' contributions. The boundaries between the inner and outer circles are fluid. New members often start off in the outer circle, keeping a close eye on the activities of the inner circle before joining it by making their own contributions. Participation in the inner and outer circles may also depend on the particular sub-issue that is currently being dealt with.

Cooperation system or organisation: CoPs that are mainstreamed in an organisation or cooperation system develop new, relevant knowledge, thereby supporting the cooperation system or the organisation as a whole. This means that the cooperation system or organisation can benefit directly from the CoP's products if all members of the cooperation system or organisation have read access to the CoP's online platform. Experience shows, however, that CoPs often need protected spaces for discussions to ensure close cooperation and build trust.

Step 3: Support the development of a CoP

The following checklist will help you support the development of CoPs. It will also help you to routinely assess the current situation and agree on the CoP's future development or disband it where necessary.

Criteria	Key questions
Purpose	■ Is the topic strategically relevant for the cooperation system or organisation? ■ Are the members genuinely interested in the topic? Does the topic reflect their needs?
Composition of the CoP	■ Do all members have adequate practical experience? Is their experience sufficiently broad? ■ Are enough individuals with specialist expertise involved? ■ Is the diversity of members guaranteed? (representatives of different fields of activities, perspectives and schools of thought)
Rules and standards	■ Have responsibilities been regulated? Have common rules and objectives been negotiated? ■ Are the communication structures in line with the different needs of members? ■ Is information shared via a variety of channels? (face-to-face meetings, online platforms, conferences, workshops, etc.) ■ Is the cost-benefit ratio acceptable for members?
Structures and process	■ Does the CoP have informal, horizontal structures? Does it enable self-organisation? ■ Have the key roles been defined? (moderators, core group, inner and outer circles)
Dynamics	■ Are the members passionate about participating or do they see it as part of their routine? ■ Are regular face-to-face meetings held? Are key results emphasised and communicated? ■ Is the 'history' of the CoP passed on to new members to highlight its unique nature?
Outcomes	■ Have useful outcomes been achieved? ■ Are these being communicated to outsiders?
Resources	■ Is sufficient time available for exchanging and developing solutions? Or are members under pressure to achieve results? ■ Are members of the CoP given enough resources by their sending organisation to allow them to take part in the community? (particularly as regards working time)

Working aid 43: Checklist for CoPs

Tool 37
Organisational Diagnosis

Notes on use

Purpose	This tool will help you document information in a structured manner when providing diagnosis, assessment and consultancy services to individual organisations.
When to use it	Can be used to perform a written survey, to conduct exploratory and stakeholder interviews, within the framework of diagnostic workshops, or as a structure for evaluating comments and hypotheses.
Setting	As required by the scope and objectives of the organisational diagnosis.
Facilities and materials	You should prepare the questionnaires/printouts of the matrix of aspects as required.
Notes	You will require a basic knowledge of organisational consultancy. The tool described here is the short version of a toolkit for carrying out an organisational diagnosis.

Description

You carry out an organisational diagnosis to assess the effectiveness of an organisation and/or as part of an organisation's change process.

An organisational diagnosis gathers information on the following aspects within an organisation:

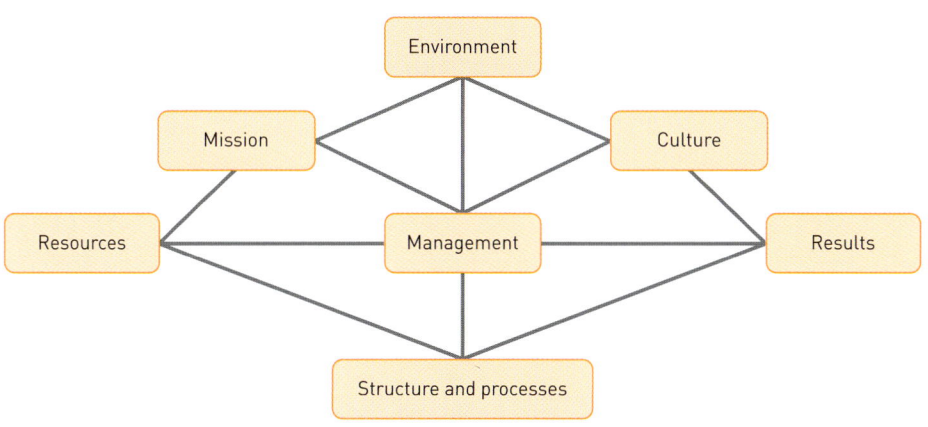

Figure 44: Matrix of aspects for organisational diagnosis

How to proceed

Step 1: Define the remit and plan the diagnosis

Start by clarifying the exact remit. This involves a process of negotiation between key individuals at the relevant organisation (or client) and the individuals conducting the diagnosis.

The starting point for any organisational diagnosis is a **clear definition of the objective**. Conflict, inefficiency, ineffective performance, weak decision-making and communication structures are all indicators that organisational structures and processes are not working as they should. An organisational diagnosis sets out to identify the reasons behind these problems and establish how you can address them. It is usually the first step in triggering a comprehensive change process within an organisation.

Defining the scope of the object under investigation goes hand-in-hand with identifying the objective. Although organisational diagnosis is geared first and foremost to individual organisations in their entirety (profit organisations, non-profit organisations, governmental organisations, institutions), it can also be used for organisational units – i.e. clearly defined parts of an organisation – provided that the issues addressed are modified accordingly.

Ideally, you should discuss and clarify the **general conditions** in a face-to-face meeting. Address the organisation's interests and expectations of the diagnosis team, the team's composition, and the support to be provided by the managers and staff members within the organisation. It is also important for the organisation to appoint a decision-making body.

Reach agreements on the following:

- How will the diagnosis proceed? What steps are planned? What survey methods will be used? Who will be surveyed?
- How much scope will the diagnosis team have? How much scope will the members of the organisation supporting them have?
- Within what time frame will the organisational diagnosis be conducted?
- How will the diagnostic process be coordinated within the organisation?
- What are the anticipated costs (in terms of time, infrastructure, resources)?
- Will the information provided by the members of the organisation be treated confidentially?
- Who will have access to the survey material?
- What will happen to the findings?

Step 2: Implement the organisational diagnosis

Depending on the methodological approach used for the diagnosis, during the implementation phase interviews are conducted, workshops are held and surveys organised. While you can conduct some of the activities of the preparatory phase remotely, the activities of the implementation phase usually require the diagnosis team to be present on the ground. Use questionnaires to help you rate the different aspects and sub-aspects during the implementation phase of the organisational diagnosis.

The seven items summarised in the following working aid are only suitable for a very general organisational diagnosis.

Questions on the different aspects	Rating and action required					
Environment Is the organisation's position within its radius of influence (market) clear to the outside world?	strongly agree ☐	agree ☐	neither agree or disagree ☐	disagree ☐	strongly disagree ☐	action required ☐
Mandate Does the organisation have a clear definition of its mandate and mission?	strongly agree ☐	agree ☐	neither agree or disagree ☐	disagree ☐	strongly disagree ☐	action required ☐
Management Does the organisation have well-functioning management structures in place that give due consideration to the challenges it faces?	strongly agree ☐	agree ☐	neither agree or disagree ☐	disagree ☐	strongly disagree ☐	action required ☐
Resources Does the organisation have sufficient resources (staff, infrastructure, etc.) to fulfil its mandate?	strongly agree ☐	agree ☐	neither agree or disagree ☐	disagree ☐	strongly disagree ☐	action required ☐
Culture Is the organisation committed to clearly defined basic values that guide its actions?	strongly agree ☐	agree ☐	neither agree or disagree ☐	disagree ☐	strongly disagree ☐	action required ☐
Structure and processes Are the organisational structures and processes for the division of labour easy to understand?	strongly agree ☐	agree ☐	neither agree or disagree ☐	disagree ☐	strongly disagree ☐	action required ☐
Results Is the organisation able to meet client expectations with the means available to it?	strongly agree ☐	agree ☐	neither agree or disagree ☐	disagree ☐	strongly disagree ☐	action required ☐

Working aid 44: Rating table for organisational diagnosis

You can break down these aspects even further by asking questions on the following keywords:

Environment

- positioning of the organisation
- clients/recipients of outputs
- competitors/peer organisations
- relationships/partnerships
- other stakeholders
- society

Mandate

- mission and purpose of the organisation
- range of products/outputs
- system of objectives (vision, goals, strategies)
- interests, motivation

Management

- management and leadership (autonomy, style of management/leadership, etc.)
- information and steering systems (costing, controlling, budgets, etc.)
- feedback and incentive systems (recognition, rewards, sanctions)
- communication and monitoring systems.

Resources

- staff and managers
- knowledge and expertise
- machinery, plant, procedures, technologies
- means of production, infrastructure
- materials, raw materials, sources of supply
- financial resources

Culture

- values and norms (client orientation, HR policy, sustainability policy, etc.)
- behavioural patterns and rules

- structures and relations of power
- definition of success
- history of the organisation

Structure and processes

- career structures, division of labour
- allocation structure for decision-making competences
- processes
- coordination
- documentation and filing

Outcomes

- operational success
- financial success
- satisfaction of stakeholders
- innovations, capacity development to consolidate success

Step 3: Evaluate findings and make recommendations

During the evaluation phase, you systematically analyse and reflect on the information collected by the diagnosis team. The evaluation primarily involves forming hypotheses, and drawing conclusions from the gathered information. Based on the need for action diagnosed, you draft recommendations for activities that are designed to improve the relevant core aspect.

Step 4: Provide feedback on findings and recommendations

In the feedback phase, you put the insights gained in the form of presentations and reports, and pass them on to the relevant key persons. Ideally, this will take the form of feedback in dialogue.

We also recommend that you hold a diagnostic debriefing session or sessions with those responsible in order to evaluate which parts of the organisational diagnosis went well and which ones did not.

The organisational diagnosis does not include the drafting of a detailed change strategy for the entire organisation or for individual parts of it. You may, however, compile such a strategy as a next step as part of a standalone change project for the organisation under review.

Tool 38
Quality Management in Organisations

Notes on use

Purpose	You can use this tool to systematically analyse strengths and weaknesses and to develop change processes for organisations or organisational units. It will provide you with a basis for planning and steering change processes.
When to use it	You can use this tool in two ways: ■ selectively when there is a specific need to improve process flows, for example. ■ routinely for ongoing learning and change processes within an organisation. One example would be analysing the performance capacity of the ministry of agriculture in a resource conservation project.
Setting	Between five and ten staff members for conducting a quality management workshop on individual topics; in the management team, to assess comprehensively the organisation's current situation.
Facilities and materials	A workshop tool case, flip chart, pinboards and the specification and assessment sheets.
Notes	Good quality management skills will be an advantage when structuring the process (e. g. Total Quality Management, European Foundation for Quality Management – EFQM).

Description

Whether you use it selectively or on a regular basis, the tool 'Quality management in organisations' makes effective use of untapped resources. It leverages potential by actively involving staff members in organisational development and getting them to exchange information and different perspectives on the organisation. This tool will help you to actively tap into existing know-how within the organisation and to implement or support targeted change processes effectively and in a structured manner.

Staff members use prescribed descriptions to assess the quality of individual aspects of the organisation or organisational unit. A brief, one-page document known as a quality specification forms the key component of this self-assessment procedure. This document lays down the quality requirements for the topic in hand, assigning them to five different stages. Figure 45 depicts the five stages of an analytical structure for the topic of 'meetings':

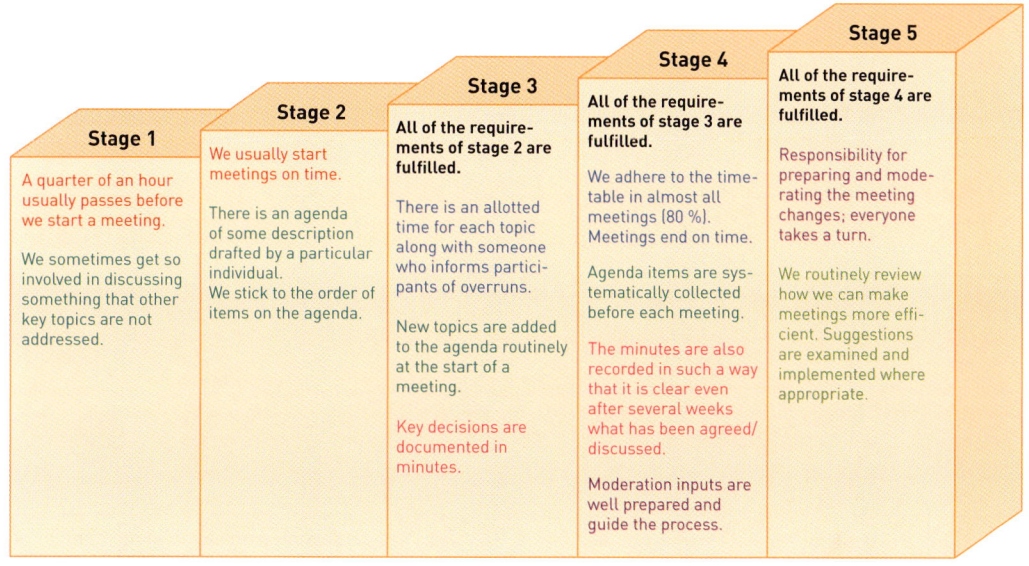

Figure 45: Example of aspects of quality management

The individual aspects of quality management (depicted here using different colours) are developed further in each stage. Some aspects appear in stage 1 and are developed until they are fulfilled in a subsequent stage. There is no specific mention of them in the next phase. 'All of the requirements of stage x are fulfilled' indicates that development has been completed. Other aspects only appear at a more advanced stage.

You can use the quality model of the European Foundation for Quality Management (EFQM) to draft suitable specifications[32]. The EFQM quality management model 'Quality as a process' (QaP)[33] was developed specifically for the field of international and development cooperation. It comprises a list of topics, along with 'best practice' procedures for dealing with them. A process that complies with QaP is eligible for certification. The model is based on the nine EFQM criteria with 32 sub-criteria for enablers in the areas of leadership, strategy, people, partnerships and resources, processes/products/services and for results in the areas of people, customers, society and business.

Two possible applications are described below.

- **Variant A** (selective use) involves conducting quality management workshops with the relevant individuals based on a specific need for action. On this basis, you develop selective change processes to improve certain procedures, structures, services, etc.

- In **variant B** (routine use), quality management procedures are implemented on an ongoing basis to drive comprehensive learning and change processes within the organisation.

How to proceed

Variant A. Selective use: Implement quality workshops

Schritt 1: Bestimmung des Handlungsbedarfs
Anlass, Ziel und Kontext des Qualitätsworkshops werden zunächst festgelegt.

Step 1: Determine the need for action
Start by identifying the reason for conducting a quality workshop, its objective and the context.

Step 2: Identify the participants
Who are the knowledge bearers for the topic to be addressed? Do you need to incorporate different perspectives (e. g. of the providers and recipients of internal services, management/staff etc.)?

Step 3: Identify a moderator for the quality workshop
Who has a neutral stance as regards the topic to be addressed? Who can moderate quality workshops?

Step 4: Select a topic from the list
Is there a specification in the list of topics that matches the need for action? What aspects contained in the specification are particularly important for the particular action required? What terms may need to be clarified in the workshop?

Step 5: Prepare quality management workshop
To prepare the workshop's content, ask yourself 'What topic is being addressed?' and put it into the organisational context ('Why is the topic important for us?').

You also need to make organisational arrangements such as selecting a suitable venue for the workshop, sending out invitations to participants on time, organising materials such as a workshop tool case, flip chart, pinboards and special work materials (specification, assessment sheets).

Step 6: Conduct the quality management workshop
Schedule two hours for the workshop. The moderator guides the group through the process. Staff members assess quality by rating specific elements and providing verification, identifying strengths and weaknesses and formulating specific selective change processes. The group then discusses these individual assessments so that the 'big picture' emerges as the workshop progresses. At the end of the workshop, the participants reach a consensus on one or two processes that describe how exactly the organisation can improve in the area under review. The moderator writes down pointers on the key stages of the workshop on a flip chart for everyone to see, most importantly on the change processes agreed at the end.

Step 7: Implement and monitor the selective change processes
Implement the agreed processes once they have been approved by the responsible individuals. Communicate the relevant information throughout the organisation or organisational unit. Where possible, implement the relevant activities within six months in order to tap into energy for change. For the purposes of transparency, use committees and other appropriate channels within the organisation to report on progress.

Variant B. Routine use:
Implementation of learning and change processes within an organisation

Step 1: Identify the objective of the quality management process and select personnel
Identify the reason and context for quality management as well as the objective of the quality management process. Nominate a representative quality management/steering group that is responsible for implementation and issue its mandate.

Step 2: Conduct an assessment of the organisation's current situation
The quality or steering group assess where the organisation 'is at' based on the nine criteria of the EFQM model. For this self-assessment, the group uses a list of topics to rate the quality of its organisation using 32 criteria (specification). The members of the group rate the criteria individually before discussing them together. After the discussion, each member rates the criteria anonymously a second time. The average value of this second rating forms the basis for the assessment. By conducting the assessment again at a later stage, you can compare the results to identify whether your organisation has made any progress ('organisational development'). This assessment only ascertains strengths and weaknesses. Individual ratings are discussed and a final second rating carried out. Unlike variant A, where quality management workshops are carried out, no change processes are developed here. Instead, analysing the assessment results over time will give you an insight into where changes need to be made. It usually takes one-and-a-half days to assess an entire organisation's current situation.

Step 3: Plan the change process
Select change topics: the quality management/steering group decides whether the quality achieved in the individual topics is sufficient or whether change processes are a) required and b) justified (i.e. is the effort involved reasonable and proportionate?). Key issues here include: importance of the topic, motivation for participants to change; in what area can a lot be achieved with little effort? Now describe the need for action in the areas selected along with the intended results that the suggested changes will bring about. For each topic, determine how and by when change processes will be developed, and who will be responsible for implementation. Carry out this step directly after assessing your organisation's current situation. You will need about a half a day.

Step 4: Develop change processes
Here, you develop change processes in quality management workshops, in line with the plan devised in the previous step. (For more details, see: Variant A Selective use, steps 5 and 6). Depending on the topic you need to address, quality management workshops can be conducted either in the quality management/steering group or in groups with other participants. You will need no more than two hours for each workshop.

Step 5: Make a decision and draft a plan of operations
Here, the quality management/steering group views the proposed change processes, prioritises them and compiles an overview. A separate implementation plan is compiled for complex processes. You will need up to one hour to carry out this step.

Step 6: Implement and monitor the change process
In this step, you communicate the plan of operations and the change processes to the workforce. The line manager or the corresponding unit reviews implementation of the processes, the achievement of objectives and the results on an ongoing basis and establishes whether the organisation is on track to achieve the intended results. Communicating information about progress to staff members will increase transparency and increase their motivation to become actively involved.

Time frame:

The time required to implement quality management procedures will depend on the complexity of the individual change topics. To sustain momentum, the process should take no longer than between ten and twelve months.

We recommend that you bundle several small change packages together so that staff members can see the improvements and experience success. If the scope of the project is too vast, participants often lose touch with why the changes were being implemented in the first place. Once you have implemented these quality management procedures, it makes sense for the steering group to assess the organisation's current situation again.

Positive side effects:

In addition to measuring quality throughout the entire organisation, assessing where your organisation 'is at' will help support team building and knowledge management. It will also allow new staff members and line managers in particular to gain a rapid and comprehensive overview of the organisation.

Tool 39

Quality Assurance in Competence Building

Notes on use

Purpose	This tool will help you improve learning results at the individual level as measured using basic didactic principles. It will also help you to dovetail development and learning objectives.
When to use it	For purposes of quality assurance: Do the planned activities take sufficient account of learning results?
Setting	Either in a small group or for purposes of individual reflection.
Facilities and materials	Sufficient number of copies of the seven basic didactic principles and of the four areas of competence.
Notes	You need to define learning objectives for a project in addition to sectoral objectives (cf. the tools 'Developing learning objectives' and 'Reviewing a project learning strategy').

Description

You can use this tool to review the learning results of planned activities at the individual level. It will help you develop competence building and quality assurance activities.

Competence building is based on a continuous process of further developing knowledge and skills. Suitable activities are geared towards the context of the learners themselves. The activities link into how the learners see themselves and foster their ability to engage in and assume ownership of self-managed learning. The focus here is on developing forward-looking competences that expand the learner's ability to reflect and take appropriate action.

How to proceed

Step 1: Review compliance with didactic principles

Start off by reviewing whether or how well a planned competence development activity complies with the seven didactic principles for improving learning results.

- **Ownership and self-organisation**
  Is the activity designed to enable self-steering and to foster self-organisation and ownership? Do the learners have the scope to independently shape the learning process? Are they involved in determining learning objectives and selecting learning methods? In other words, does the activity help and encourage the learners to assume ownership of the actual learning process, and consolidate and further develop what they have learned?

- **Learning support/advice**
  Is systematic support and advice in place that allows learners to reflect on their own competence profile? Are individual learning strategies and learning processes analysed? Are appropriate learning speeds and learning pathways identified? Is technical expertise made available?

- **Multiple perspectives and switching of perspectives**
  Are multiple perspectives guaranteed? Are the learners able to switch perspectives? Is the learning group heterogeneous, enabling multiple perspectives and switching of perspectives? Are opinions and patterns of action viewed from a different perspective and questioned? Are additional solutions and options sought?

- **Attitudes**
  Is an attitude of mutual regard and respectful comparison fostered?

- **Spaces of experience**
  Are learning spaces designed to enable experimentation and reflection? Do they enable learners to undergo new experiences and see themselves and others in a new light?

- **Reflection**
  Is the effectiveness of each learner's own actions and of joint actions viewed in a critical light? Is enough time devoted to individual self-reflection and joint deliberations? Does the reflection process allow the close examination and adjustment of learning and of its effectiveness?

- **Co-construction of learning**
  Are relationships, dialogue, communication and cooperation designed so that learners can develop innovative solutions together?

Step 2: Review compliance with areas of competence

In this second step, you review the learning process to establish whether the following areas of competence are being addressed, and if so, how. Are they relevant for achieving the project objectives?

- **Sectoral competence**
  Does the learning process foster sectoral know-how, enabling learners to reflect on the specific problems in the project in an appropriate manner and develop possible options for achieving the development objectives?

- **Methodological competence**
  Does the learning process promote methodological skills that are geared to addressing the challenges in the area of social concern?

- **Social competence**
  Does the learning process foster social expertise that will support cooperation and communication among actors in the area of social concern?

- **Personal competence**
  Does the learning process support personal skills that will enable the relevant individuals to participate actively in the area of social concern?

The following questions will help you shape learning processes:

- Does the learning content link into the context and working day of the learners? Are the learning processes geared towards the actual situation in which learners will put what they have learned into practice?

- Are appropriate forms of learning used? Do these provide learners with optimal access to knowledge and opportunities to try things out?

- Is learning support structured so as to enable learners to reflect on and apply what they have learned?

Step 3: Review the effects of competence building on other levels of capacity development

Here, the focus is on examining learning processes at the personal level to determine how appropriate they are for fulfilling the objectives in the area of social concern, in addition to attaining individual learning goals. To do this, you describe the results that activities at the individual level will have on the organisational and societal levels.

The following questions will help you design competence building activities:

- Is the learning process designed to achieve results beyond the individual level?

- What influence will the learners have on other important groups of actors?

- Are the learners institutionally integrated to a sufficient degree?

- Will the learning process enable participants to help initiate, implement and manage intended change processes? Do the competence development activities include specific steps to ensure that participants can make this contribution, such as mainstreaming through transfer projects, prototyping, communities of practice, training of multipliers etc.?

- Are learners able to make an active contribution to change processes (motivation, existing knowledge, change management skills and leadership expertise)?

- Have multipliers received suitable training and capacity building?

- Can the learning results be scaled up?

- Do learning networks (peer-to-peer networks for sharing practical experience, alumni work) and suitable follow-up activities support the sustainability of learning results?

Step 4: Define activities to improve learning results

If starting points for optimising learning results were identified in steps 1, 2 and 3 above, in this step you can consider together with the other actors involved how to further improve the learning results of the competence building activities.

Together with the other actors, reflect on the following questions, which will help you to optimise learning results:

- Do the identified needs reflect the priorities of the learners?

- Is the learning content used in this context relevant and context-oriented?

- Have suitable participants been identified?

- Is the composition of the learning group appropriate?

- Does the learning content match the participants' current level of knowledge and their working day?

- Are suitable media and forms of learning used?

- Is learning support and the role it plays appropriate to the learning content and the participants involved?

- Has enough scope been created for reflection and ownership of the learning process by the learners?

Tool 40
Intervision

Notes on use

Purpose	Intervision is a structured group meeting in which one participant is advised by the other members – who have been cast in particular roles. The aim is to develop joint practical solutions for a specific problem.
When to use it	In situations where you need to address the following themes, for example: ▪ tackling new tasks; ▪ cooperation within groups and between organisations; ▪ dealing with unfamiliar behaviours; ▪ integrating new staff members; ▪ breakdowns in working procedure; ▪ difficulties with superiors.
Setting	The ideal group size is between six and nine. If the number of interested parties is larger, it is a good idea to form several groups. Intervision takes about one hour.
Facilities and materials	Flip chart.
Notes	The 'storyteller' should bring along a clearly formulated issue or problem. You should make an effort to maintain a constructive and cooperative atmosphere, thereby ensuring that you support the storyteller by communicating effectively and providing helpful information.

Description

Intervision – or peer-to-peer learning – is an uncomplicated method for using implicit knowledge. It harnesses the experience that people with similar professional backgrounds can use to advise each other.

Intervision has a specific structure. It is carried out among peers. The composition of the group, and the fact that it is self-steering, promote this process of transforming implicit knowledge into explicit knowledge by ensuring openness and a practical orientation.

You should also apply the following principles in intervision:

▪ Intervision is conducted by a group on a self-reliant basis, i. e. without an external moderator. This takes place within a structured framework in which the roles are cast and the communication process is controlled

- The fact that intervision involves a peer group by definition means that there are no hierarchy gaps, and that peers can express themselves openly without feeling inhibited by the presence of superiors.

- The groups itself decides when and where it wishes to meet, normally at intervals of between two and six weeks.

- The group can also establish a virtual intranet/internet-based platform (collaborative work group) in order to coordinate its activities and record the outcomes. The group itself decides when and to whom this information should be made available. These platforms do not replace face-to-face meetings.

- At the first session, participants discuss the rules of intervision, and reach various agreements within the group. During the first part of the session, the method is explained by an individual who is already familiar with it.

How to proceed

Step 1: Allocate roles

At each session, participants are assigned the following roles: storyteller, moderator, timekeeper, advisor (of which there are several) and observer.

Step 2: Tell the story and define the key question

In this step, the storyteller briefly describes the situation as he/she sees it. The moderator jots down key points that summarise the key statements regarding the situation. Together with the storyteller, he/she agrees on key questions that intervision will help address. The moderator makes a note of the main question.

This step should take no longer than between five and ten minutes.

Step 3: Ask questions about the story

Here, the advisors ask questions on the context and the specific background or on a particular perspective in order to gain a deeper understanding of the story. The storyteller then answers these questions in the necessary degree of detail.

This sequence should also take no longer than between five and ten minutes.

Step 4: Compile hypotheses on the situation

The advisors indicate what struck them most about the story and offer hypotheses (assumptions) about the situation ('What is going on here'?). The hypotheses are not discussed or assessed. The moderator writes down the hypotheses on a flip chart. The storyteller simply listens.

This sequence can take about ten minutes.

Step 5: Comment on the hypotheses

In this step, the storyteller comments on the hypotheses compiled, selecting one or two that he/she finds particularly appealing. The storyteller provides more information if he/she feels that this will help generate ideas to address the issue in hand.

This step should take no longer than five minutes.

Step 6: Put forward solutions

For the hypotheses selected, the advisors suggest ideas on how to proceed, as well as possible solutions that the storyteller could use. (What would you recommend? What solutions have proven successful in a similar situation?) The moderator writes down the ideas on a flip chart. Again, the storyteller simply listens.

This sequence can take about ten minutes.

Step 7: Comment on proposed solutions

In this step, the storyteller comments on the ideas which he/she thinks are helpful at first glance, and states what he or she will be taking away from the intervision.

This step should take no longer than five minutes.

Step 8: Evaluate ideas at the meta level

The moderator takes brief feedback. The observer gives feedback to all the participants. Participants reflect on the process as a group. (Note: do not dwell on matters too much!)

Once again, you should allow about five minutes for this step.

Tool 41
Developing Learning Objectives

Notes on use

Purpose	This tool will help you to reflect on the ability of a cooperation system to learn and change and to devise specific learning objectives at the three levels of capacity development – individuals, organisations and society (including the elements of cooperation systems and enabling frameworks). It includes an analysis of strengths and weaknesses based on the three mechanisms of learning (variation, selection and stabilisation).
When to use it	Suitable for developing and differentiating between learning objectives. In-depth analysis of which Capacity WORKS tools you can use to achieve which learning objectives.
Setting	Workshop with key actors.
Facilities and materials	Possibly transfer tables to flip chart or pinboard.
Notes	This tool is demanding in that the participants are expected to have an in-depth understanding of the model, and especially the three learning mechanisms (variation, selection and stabilisation).

Description

This tool will help you reflect on learning within the project based on aspects such as:

- … the five success factors of the Capacity WORKS management model;
- … the three levels of capacity development (individuals, organisations, society);
- … the three mechanisms of learning (variation, selection and stabilisation).

How to proceed

Step 1: Assess the ability for learning and change

In this first step, you analyse the strengths and weaknesses of the project in relation to its ability to implement the three learning mechanisms (variation, selection and stabilisation). For each success factor, you conduct an analysis of how well variation, selection and stabilisation are being incorporated into the achievement of the project's objectives.

It will help if you fill out the following working aid for each of the five success factors and assess the strengths and weaknesses based on specific practical examples or experiences from the project.

Success factor: ...	Strengths	Weaknesses
Variation	Assessment: Examples:	Assessment: Examples:
Selection	Assessment: Examples:	Assessment: Examples:
Stabilisation	Assessment: Examples:	Assessment: Examples:

Working aid 45: Analysis of strengths and weaknesses based on the three learning mechanisms

The following overview will help you identify deficits and conversely, highlight the respective strengths.

Indications of deficits in variation:

- too little information, imagination and experience; too few ideas;
- too much routine;
- formation of stable, entrenched positions;
- formation of rigid, entrenched positions.

Indications of deficits in selection:

- too much information, experience and imagination; too many ideas;
- weak decision-making;
- unclear decision-making structures;
- avoidance of positive selection.

Indications of deficits in (re-)stabilisation

- a lack of rules;
- too little continuity;
- a lack of orientation;
- a lack of commitment;
- weak implementation capacity;

- broad-based lack of acceptance;

- a lack of resilience (selection of volatile rather than resilient solutions);

- too little compliance.

Remember that this issue links learning objectives very closely indeed with project objectives. Thus the issues at stake are, for example, variation in strategy formation, selection in processes, or stabilisation in steering.

Step 2: Define learning objectives

You now define learning objectives based on the analysis of strengths and weaknesses. These learning objectives supplement the project objectives by highlighting the learning aspects that are intended in the project objectives, frequently at an implicit level.

Step 3: Assign learning objectives to the levels of capacity development

In this step, based on the analysis of strengths and weaknesses, you define the project's learning objectives at the three levels of capacity development – individuals, organisations and society (including the elements of cooperation systems and the development of enabling frameworks). You assign them, and prioritise them where necessary (e.g. using stickers). The following table will help you do this:

Learning objectives by level	Variation	Selection	Stabilisation
Individuals			
Organisations			
Society: cooperation systems			
Society: enabling frameworks			

Working aid 46: Identifying learning objectives

Step 4: Develop indicators

Here, you can think about how you would determine whether and/or how well the learning objectives have been achieved for each level. You can enter your thoughts on this in the following table:

How will you recognise whether you have achieved the defined learning objectives?	
Individuals	
Organisations	
Society: cooperation systems	
Society: enabling frameworks	

Working aid 47: Indicators for learning objectives

Step 5: Select appropriate tools for learning development

Which tools can support variation, selection and stabilisation in the prioritised objectives/activities?

Figure 46 will provide you with some guidance. It shows you which tools in Capacity WORKS promote which of the three mechanisms of variation, selection and (re-)stabilisation.

When the tools are used flexibly, these relationships may change.

You can use some of the tools for more than one mechanism, in which case the numbers for the tools appear on the line between two boxes.

Learning mechanisms \ Tools	SF Strategy	SF Cooperation	SF Steering structure	SF Processes	SF Learning & innovation
Variation	2, 3, 5	10, 11, 13			32, 35, 36, 40
	1, 4	9, 12, 17			39
Selection	6, 7, 8		20	25	41
		14, 19	23, 24	27	30, 37, 38, 42
(Re-)stabilisation		15, 16, 18	21, 22	26, 28, 29	31, 33, 34

The numbers correspond to the numbers of the tools in the respective success factors.
SF = success factor

Figure 46: Assigning the Capacity WORKS tools to the three learning mechanisms

Tool 42

Reviewing a Project Learning Strategy

Notes on use

Purpose	This tool will help you review, coordinate and, where necessary, modify a project's activities with respect to their contribution to learning at all three levels of capacity development – individuals, organisations and society (cooperation systems and enabling frameworks) – based on the capacity development strategy.
When to use it	To check the consistency of hypotheses used as a basis for the project strategy. Can be used if implementation of the strategy becomes deadlocked or if the strategy is regarded as inadequate for achieving the targeted capacities.
Setting	Workshop with key actors.
Facilities and materials	You may need to transfer the table to a flip chart or pinboard; otherwise provide copies of work tables.
Notes	Requirements: The project already has a capacity development strategy in place and the participating actors must be familiar with it.

Description

This tool will help you identify, review and where appropriate prompt optimisation of the learning strategy pursued (either implicitly or explicitly) in any project. A fully developed capacity development strategy provides a starting point for the joint project.

How to proceed

Working aid 48 breaks down the review of the learning strategy into six steps. Use it to guide yourself through the process:

(Selected elements of) project XY: Review of the learning strategy				
	Individuals	Organisa-tions	Society	
			Cooperation systems	Enabling frameworks
Step 1: What learning objectives are targeted at the different levels? (What are the intended capacities)?				
Step 2: What activities and associated hypotheses have already been planned?				
Step 3: Are the activities suitable for reaching the learning objectives?				
Step 4: To what extent do the planned activities re-inforce each other as regards the achieve-ment of learning objectives at the level in question? (Potential for synergies)				
Step 5: To what extent do the planned activities ob-struct each other as regards the achieve-ment of learning objectives at the different levels? (Potential for conflict)				
Step 6: What changes/enhancements to the planned activities do the previous steps trigger?				

Working aid 48: Review of the learning strategy

Step 1: Identify learning objectives at the different levels

Here, you define the focus for the analysis: which particular elements of the project will you re-view? To identify this focus, describe the learning objectives at the three levels of capacity devel-opment, which can be derived from the planned project results (cf. the 'Intended capacities' line in the tool 'Capacity development strategy').

Step 2: Describe planned activities

In this second step, you transfer the activities that have already been planned at all levels of capac-ity development along with the corresponding hypotheses.

Step 3: Review activities with regard to the learning objectives

In step 3, you check whether the planned activities are the right 'fit' for the targeted learning objectives. Ideally, the activities should be able to achieve the learning objectives at all points of contact and at the three levels of capacity development.

Step 4: Identify positive synergies

Here, you check whether positive synergies are generated between the planned activities. To what extent are the activities mutually reinforcing as regards the achievement of learning objectives at the three levels of capacity development?

Step 5: Identify potential for conflict

In this step, you check the extent to which the different activities could give rise to conflict as regards the achievement of learning objectives at all three levels.

Step 6: Draw conclusions

In this final step you assess the interim results achieved in the previous steps and identify any suggestions for changing or enhancing the planned activities in order to achieve the desired learning objectives.

Citations

1 Paraphrased from Wimmer, Rudolf (2009): Führung und Organisation – zwei Seiten ein und derselben Medaille, in: Revue für postheroisches Management, Heft 4, 2009, pp. 20–33.

2 Cf. König, Helmut (2008): Politik und Gedächtnis, Weilerswist, pp. 368ff..

3 Cf. Abelshauser, Werner (2005): Deutsche Wirtschaftsgeschichte seit 1945, Bonn, pp. 75ff.

4 Mintzberg, Henry (1978): Patterns of Strategy Formulation, in: Management Science 24, pp. 934–948.

5 Cf. Nagel, Reinhart; Wimmer, Rudolf (2009): Systemische Strategieentwicklung. Modelle und Instrumente für Berater und Entscheider, 5. akt. und erw. Auflage, Stuttgart, pp. 25–100.

6 Cf. Ulrich, Hans; Krieg, Walter K. (1972): Das St. Galler Management Modell, Bern.

7 Based on Nagel, Reinhart; Wimmer, Rudolf (2009): Systemische Strategieentwicklung. Modelle und Instrumente für Berater und Entscheider, 5. akt. und erw. Auflage, Stuttgart, pp. 103ff.

8 Nagel, Reinhart; Wimmer, Rudolf (2009): Systemische Strategieentwicklung. Modelle und Instrumente für Berater und Entscheider, 5. akt. und erw. Auflage, Stuttgart, pp. 75.

9 Based on Payer, H. (2011): Organisation, Kooperation, Netzwerk – Fließende Übergänge zwischen fester und loser Kopplung, in: Ahlers-Niemann, A., Freitag-Becker, E. (eds.): Netzwerke – Begegnungen auf Zeit zwischen Uns und Ich, Bergisch Gladbach, pp. 23–39.

10 Cf. Wimmer, Rudolf (2007): Die bewusste Gestaltung der eigenen Lernfähigkeit als Unternehmen, in: Tomaschek, N. (ed.): Die bewusste Organisation. Steigerung der Leistungsfähigkeit, Lebendigkeit und Innovationskraft von Unternehmen, Heidelberg, pp. 39–62.

11 Tool based on a document by Arthur Zimmermann, odcp consult GmbH, Zürich.

12 Tool developed by Arthur Zimmermann, odcp consult gmbh, Zürich. Cf. Deutsche Gesellschaft für Technische Zusammenarbeit (GTZ) GmbH (eds.), Autor Zimmermann, Arthur (n.y.), Mainstreaming Participation, Instrumente zur AkteursAnalyse, 10 Bausteine für die partizipative Gestaltung von Kooperationssystemen, Eschborn, pp. 14f.

13 Tool developed by Arthur Zimmermann, odcp consult GmbH, Zürich.

14 Basics of the tool developed by Arthur Zimmermann, odcp consult gmbh, Zürich.

15 Cf. Bauer-Wolf, Stefan, ÖAR, Vienna (2012): Ein Lebenslauf von Kooperation, published at www.coaching-raum.at.

16 After Leo Baumfeld, ÖAR, Vienna: unpublished working paper.

17 After Leo Baumfeld, ÖAR, Vienna: unpublished working paper.

18 Basics of the tool developed by Arthur Zimmermann, odcp consult gmbh, Zürich.

19 Cf. Bauer-Wolf, Stefan; Payer, Harald; Scheer, Günter (eds.) (2008): Erfolgreich durch Netzwerkkompetenz, Heidelberg.

20 Cf. Deutsche Gesellschaft für Technische Zusammenarbeit (GTZ) GmbH (eds.), Author Zimmermann, Arthur (n.y.), Mainstreaming Participation, Instrumente zur AkteursAnalyse, 10 Bausteine für die partizipative Gestaltung von Kooperationssystemen, Eschborn, pp. 32–35; cf. Nowak, Martin A. (2012): SuperCooperators: Altruism, Evolution, and Why We Need Each Other to Succeed, Free Press.

21 Covey, Stephen M. (2006): The Speed of Trust, Free Press.

22 Cf. Axelrod, Robert (2006): The Evolution of Cooperation, Basic Books.

23 Tool developed by Arthur Zimmermann, odcp consult GmbH, Zürich.

24 Tool developed by Arthur Zimmermann, odcp consult GmbH, Zürich.

25 Tool developed by Arthur Zimmermann, odcp consult GmbH, Zürich.

26 Cf. Fisher, Roger et al. (2009): Das Harvard-Konzept, Der Klassiker der Verhandlungstechnik, Frankfurt.

27 Cf. Königswieser, Roswita; Exner, Alexander (2008): Systemische Intervention, Architekturen und Designs für Berater und Veränderungsmanager, Stuttgart.

28 Graf-Götz, Friedrich; Glatz Hans (1998): Handbuch Organisation gestalten. Für Praktiker aus Profit- und Nonprofit-Unternehmen, Trainer und Berater, Weinheim, p. 152.

29 Basics of the tool developed by Arthur Zimmermann, odcp consult gmbh, Zürich.

30 Tool developed from Eppler, Martin J. (2007): Debriefing – Lernen aus Erfolgen und Fehlern, in: Zeitschrift für Organisationsentwicklung, 01/2007, Werkzeugkiste Nr. 10, pp. 73–77.

31 Basics of the tool developed by Arthur Zimmermann, odcp consult gmbh, Zürich.

32 For a description of the EFQM Excellence Model, fundamental concepts and criteria see the website of the European Foundation for Quality Management (EFQM), www.efqm.de.

33 See the website of the Swiss Association for Quality and Management Systems (SQS), www.sqs.ch.

List of figures

List of working aids

Acknowledgements

We owe our thanks to all those who made valuable contributions to the initial design and subsequent development of the model. Without them Capacity WORKS would not exist today.

During the initial phase (2004–2006) this included many project and programme managers who contributed their experience with successful project approaches to the discussion: Michael Gajo and Wolfgang Morbach, Christoph Feyen, Christopher T. Mallmann, Meinolf Spiekermann and Björn Philipp. Their partners at our Head Office were the members of the company's MODeLS consultancy team that existed at the time (Elisabeth Christian, Sylvia Glotzbach, Klaus Reiter), and its Corporate Organisation Section (Kurt Wagner, Lutz Zimmermann). Helmut Willke supported the development of the model as an external consultant. The following external consultants were involved in writing the background texts and designing the tools: Leo Baumfeld, Claudia Conrad, Jean-Pierre Wolf and Arthur Zimmermann. And without the directors general and directors of the corporate units at GTZ (as it called was at the time), Capacity WORKS certainly would not have been produced.

During the pilot phase (2007–2008), together with their respective teams and partners worldwide some 70 project and programme managers tried out Capacity WORKS, and fed their experiences back into the ongoing development process. The company's Corporate Organisation Section, which included Heiko Roehl, Kurt Wagner and Maraile Görgen, guaranteed the mainstreaming of the model through corporate policy. This included the production of a second version of Capacity WORKS together with the external consultant Mischa Skribot.

During the rollout phase (2009–2010) it was again the staff – whether national, international or seconded – who added to our experience with Capacity WORKS through their interest in trying things out, their inventiveness and their feedback during training courses, when using the model or when posting on the blog. This process was supported by the project managers at Head Office (Bernadette Daubenmerkl, Bettina De Campos, Martina Maurer, Cordula Schmüdderich, Soete Klien and Maraile Görgen) and a group of external consultants. Oliver Karkoschka and Claudia Conrad were involved in monitoring the rollout process.

This update – the third version – was written basically by the staff of the Competence Centre for Change Management & Development Approaches (Werner Ahringhoff, Elisabeth Christian, Barbara Gerhager, Sylvia Glotzbach, Joachim Göske, Neil Hatton, Soete Klien, Klaus Reiter and Joachim Stahl). Also involved were GIZ's Corporate Organisation Section (Maraile Görgen) and its Germany Department (Alexandra Pres, Balthas Seibold), as well as the external consultants Stefan Bauer-Wolf and Harald Payer.

When so many people work on a project with such commitment, it is easy to leave someone's name out by mistake. Should any of our colleagues reading this feel that their name is missing, we would ask them to accept our sincere apologies. We know that Capacity WORKS only became what it is thanks to all the contributions made.

Printing and Binding: PHOENIX PRINT GmbH, Würzburg